Undergraduate Lecture Notes in Physics

Undergraduate Lecture Notes in Physics (ULNP) publishes authoritative texts covering topics throughout pure and applied physics. Each title in the series is suitable as a basis for undergraduate instruction, typically containing practice problems, worked examples, chapter summaries, and suggestions for further reading.

ULNP titles must provide at least one of the following:

- An exceptionally clear and concise treatment of a standard undergraduate subject.
- A solid undergraduate-level introduction to a graduate, advanced, or non-standard subject.
- A novel perspective or an unusual approach to teaching a subject.

ULNP especially encourages new, original, and idiosyncratic approaches to physics teaching at the undergraduate level.

The purpose of ULNP is to provide intriguing, absorbing books that will continue to be the reader's preferred reference throughout their academic career.

Series editors

Neil Ashby
University of Colorado, Boulder, CO, USA

William Brantley
Department of Physics, Furman University, Greenville, SC, USA

Matthew Deady
Physics Program, Bard College, Annandale-on-Hudson, NY, USA

Michael Fowler
Department of Physics, University of Virginia, Charlottesville, VA, USA

Morten Hjorth-Jensen
Department of Physics, University of Oslo, Oslo, Norway

Michael Inglis
SUNY-State University of New York, Selden, NY, USA

More information about this series at http://www.springer.com/series/8917

Saverio D'Auria

Introduction to Nuclear
and Particle Physics

 Springer

Saverio D'Auria
School of Physics and Astronomy
University of Glasgow
Glasgow, United Kingdom

ISSN 2192-4791   ISSN 2192-4805   (electronic)
Undergraduate Lecture Notes in Physics
ISBN 978-3-319-93854-7   ISBN 978-3-319-93855-4   (eBook)
https://doi.org/10.1007/978-3-319-93855-4

Library of Congress Control Number: 2018956310

To Anna, Filomena and Maria-Martina

Preface

This book originates from a short introductory course given at the University of Glasgow on nuclear and particle physics. It is intended for a diverse public: students who are interested in the subject will find here some introductory material, while others may want to understand the main concepts of this branch of physics for applications in other sciences. High school teachers, who wish to refresh this subject, and, in general, the knowledgeable enthusiast may find this book useful. Its introductory nature requires some simplification and approximation. Knowing that it is much more difficult to modify wrong concepts than to acquire new ones from *tabula rasa*, an effort has been made to avoid any simplification that would require a complete change in a more advanced course. The level of this book is certainly higher than what would be considered as public understanding of science: a mathematical language is used, but at a level which should be understandable after the first year of a scientific discipline. Some familiarity with chemical symbols of elements is also assumed. For the benefit of readers with backgrounds other than physics, it was important to include some key results, which have high scientific and cultural relevance, such as Nöther's theorem and group theory. Quantum physics was not part of the course, but a dedicated chapter was included to make the book more complete. It is treated only qualitatively, at an introductory level, and in a very concise form. Nuclear and particle physics is far from being an axiomatic subject. Learning is rather a circular process, where concepts need to be anticipated to a certain extent, and they will become clearer later, because concepts may be understood at different depths. The reader is encouraged not to be afraid of new terms, which are introduced in *italic*. One or two more advanced subjects are also included. They are intended both as a stimulus to further progress in the field and as reference material in a simplified but still rigorous form. Problems are mostly case studies on real laboratory problems, and their solutions are provided separately. A final chapter with six case study problems and their solutions has been added.

In margin are some biographical notes of the physicists who have contributed to the field with discoveries and explanations. Many more have contributed with marginal, but maybe pivotal, work. Behind the discoveries are real people, from many places in the world, each with his or her own life, which was often shaped by the historical events of the last century. The international character of the research, even in a strongly divided world, is fascinating.

I would like to express my gratitude to the University of Glasgow and to Peter Bussey, Craig Buttar, Aidan Robson and Ken Smith for discussion and suggestions. Mario Spezziga, and Kate Shaw have also contributed with comments. I should also credit my teaching inspiration models: Attilio Forino and the late Giampietro Puppi, who have formed generations of students at the University of Bologna, Italy. I thank my students for asking questions, and Ali Walker for reading the manuscript, the publisher staff for their patience and my wife and daughters for patience and support. I hope the reader will enjoy reading this book as much as I did while writing it.

Glasgow, UK   Saverio D'Auria
Geneva, Switzerland
January 2018

Contents

Math Coventions Used

Greek Letters Used in Physics

α	A	Alpha	η	H	Eta	ν	N Nu	τ	T Tau
β	B	Beta	θ, ϑ	Θ	Theta	ξ	Ξ Xi	υ	Υ Upsilon
γ	Γ	Gamma	ι	I	Iota	o	O Omicron	ϕ, φ	Φ Phi
δ	Δ	Delta	$\kappa, \varkappa$	K	Kappa	π	Π Pi	χ	X Chi
ϵ	E	Epsilon	λ	Λ	Lambda	ρ	P Rho	ψ	Ψ Psi
ζ	Z	Zeta	μ	M	Mu	σ	Σ Sigma	ω	Ω Omega

Constants and Conversion Factors

Constant	Symb.	Value	Error	Units
Speed of light in vacuum	c	299792458	None	m/s
Planck's constant[1]	h	$4.135667662 \times 10^{-15}$	(25)	eV s
Vacuum permittivity	ϵ_0	8.854187817	–	F/m
Electron charge[1]	e	$1.602176634 \times 10^{-19}$	None	C
Proton mass	m_p	938.272081	(6)	MeV/c^2
		$1.672621777 \times 10^{-27}$	(74)	kg
Electron mass	m_e	0.5109989461	(31)	MeV/c^2
		$9.10938356 \times 10^{-31}$	(11)	kg
Bohr's magneton	μ_B	$5.7883818012 \times 10^{11}$	(26)	MeV T^{-1}

(continued)

Constant	Symb.	Value	Error	Units
Nuclear magneton	μ_N	$3.1524512550 \times 10^{14}$	(15)	MeV T^{-1}
Atomic mass unit	u	931.4940954	(57)	MeV/c^2
Boltzmann's constant[1]	k	8.6173303×10^{-5}	(50)	eV/K
Avogadro's number[1]	N_A	$6.02214076 \times 10^{23}$	None	mol^{-1}
Gravitational constant	G_N	6.67408×10^{-11}	(31)	m^3 kg^{-1} s^{-1}
Fermi coupling constant	$G_F/(\hbar c)^3$	1.1663787×10^{-5}	(6)	GeV^{-2}
Strong coupling constant	$\alpha_s(m_Z)$	0.1185	(6)	
Fine structure constant[2]	$\alpha = e^2/(4\pi\epsilon_0 \hbar c)$	1/137.035999074	(44)	
Electron Volt[2]	eV	$1.602176565 \times 10^{-19}$	(35)	J
barn	b	10^{-28}	–	m^2

[1] The General Conference on Weights and Measures (CGPM) has decided that effective from May 2019 the electron charge, Planck's constant, Boltzmann's constant and Avogadro's constant, together with the speed of light, are the basis to define all other International System (SI) units, and therefore are defined without any experimental error

[2] Also the fine structure constant and the electron Volt are defined with no uncertainty with the new definition of SI units

Chapter 1
Introduction to Radiation

This book is about radiation, radioactive decays and some elements of nuclear and particle physics. Einstein's theory of special relativity is needed for a correct description of some of these phenomena. The word "special" is used to distinguish this part from the theory of general relativity, that deals with non-inertial reference frames and with gravity. Some textbooks follow a rather historical, or history-driven, approach to radiation. While this approach would be quite educational, giving a view of how ideas developed almost exactly one hundred years ago, it is probably not the fastest way to explain the fundamental concepts.

Let's try to define radiation first. In general, it is any form of energy that can be emitted and sent over a distance. The word contains the root "radius", which suggests an emission of energy in all directions, like the light from the sun or a light bulb. However, radiation can be emitted by a directional antenna, or concentrated in a beam, like a laser.

Light is the form of electromagnetic radiation, which is best known by our common experience. Heat can also be radiated by a body, in the form of infra-red radiation. We are likewise familiar in radio waves, infrared (remote-control) and visible light. They are all the same electromagnetic radiation, but at different wavelengths.

The first experience of a new kind of electromagnetic radiation occurred in 1895, when Wilhelm Röntgen, in Munich, observed some fluorescence, i.e. a dim light, induced by operating vacuum tubes, which emitted what he called X-rays.

© Springer Nature Switzerland AG 2018
S. D'Auria, *Introduction to Nuclear and Particle Physics*,
Undergraduate Lecture Notes in Physics,
https://doi.org/10.1007/978-3-319-93855-4_1

Fig. 1.1 Marie Sklodowska-Curie while measuring the activity of a Radium source. Courtesy of Museé Curie, Paris. She was born in Poland in 1867 and naturalised French. She was the first woman to become professor at the University of Paris, the first woman to be awarded a Nobel prize, which she shared with her husband Pierre Curie and with Henry Becquerel; she is the only woman to be awarded two Nobel prizes in two different disciplines (Physics (1903) and Chemistry (1911))

The first radiography followed after a few weeks, and a few months later Henry Becquerel, in Paris, discovered natural radioactivity in uranium minerals: they produced a fluorescence similar to the one produced by the X-rays. Marie Sklodowska (Fig. 1.1) and her husband Pierre Curie, also in Paris, made the first quantitative measurements of radiation. Upon discovery that thorium is also radioactive, they also discovered two new radioactive elements, which they named polonium and radium. Radiation units were subsequently named after these four pioneers. The Becquerel (Bq) is equivalent to "one disintegration per second"; its physical dimensions are $[T^{-1}]$, and it measures the *activity* of a radioactive source, meaning how much radiation it produces. Sometimes, in old documents, or on old sources, another unit is used, known as the Curie (Ci) where 1 Ci $= 3.7 \times 10^{10}$ disintegrations per second (Bq), 1 Ci$= 37$ GBq. Historically, it is roughly the activity of 1 g of ^{226}Ra. This is a non-SI unit, no longer in use. The unit named after Röntgen similarly became obsolete.

A little digression on the units: although we can always use SI units, it is more convenient to use units that are suited to the quantity that we are going to measure. In Astronomy, we can use *light years* and *parsec* for distances; for extremely small systems, we use the *electron Volt (eV)* as the unit of energy. One *eV* is the kinetic energy gained by an electron which is accelerated by an electric potential difference of 1 V. One *Joule* is one *Coulomb* × *Volt*. The electron charge is $1.602176565(35) \times 10^{-19}$ C, so 1 eV $= 1.602176565(35) \times 10^{-19}$ J. We should note at this point that only the electron charge enters in this definition, not its mass. As the electron and proton electrical charges, in absolute value, are experimentally equal to each other, at most, one part in 10^{18}, the unit could as well have been called "proton-Volt".

We can classify radiation based on its source: natural radiation from minerals in the Earth's crust is the most common. Radon gas is a common source of background to many precision physics experiments and the main source of radiation dose to the population in mainland Europe and the USA. The presence of Radon in the air, especially in basements, can produce an activity up to 120 Bq per cubic metre of air. Within Britain, Cornwall and Wales have the largest background activity from Radon, up to 30 Bq/m^3. Cosmic rays, spanning from the so-called *solar wind*, to high-energy cosmic rays, are also a large natural source of radiation. After "showering" through the atmosphere, in the terrestrial magnetic field, cosmic rays reach the ground and are a source of natural radioactivity. Cosmic rays also produce radioactive elements (*nuclides*) in the atmosphere: for instance, ^{14}C.

Radon is a common noble gas, which is continually formed by decay of radioactive elements. It is unstable and decays

$$^{222}\text{Rn} \rightarrow \alpha + {}^{218}\text{Po} \rightarrow \alpha + {}^{214}\text{Pb} \rightarrow \ldots$$

This formula will be explained later. In areas where Radon is an issue, ventilating the basements of houses helps reducing the health risk associated.

Examples of man-made sources of radiation include radioactive nuclides, which are produced intentionally for medical use, or as a side product in nuclear power plants and nuclear explosions; X-ray diagnostic devices and particle accelerators, which can also produce "synchrotron radiation", a coherent high-energy X-ray beam. The Cherenkov radiation is just light, which is emitted by charged particles when traveling in a homogeneous medium at a speed which is larger than the speed of light in that medium. Cosmic rays emit this radiation when propagating in the atmosphere. A *transition radiation* is emitted when fast charged particles cross the interface between different materials.

Radiation can be classified also according to its effects. Some types of radiation ionise the medium they traverse: all the electrically charged particles, such as alpha and beta particles, which we'll be introduced to later, ionise the medium they traverse and lose energy. Also, electromagnetic radiation, such as gamma and X-rays, ionise materials. Neutron radiation does not directly ionise, but interacts in a different way, by elastic and quasi-elastic scattering, or by inducing nuclear reactions like fission. We'll see all this later in this book. We assume in the following that the reader is familiar with the atomic structure of matter and with the fact that atomic nuclei are made of protons and neutrons. Protons are positively charged particles; their charge is exactly the same as the charge of electrons, with opposite sign, but their mass is 1860 times larger than the electron mass. Neutrons have no electrical charge, and their mass is approximately equal to the proton mass, just slightly larger.

The properties of a chemical element are defined by the structure of the electron cloud surrounding the nucleus. Only the number of protons in the nucleus defines the number of electrons surrounding it and, hence, the chemical properties of the atom. When we generically refer to carbon, we specify an atom made with a nucleus with six protons surrounded by six electrons. However, we do not specify how heavy the nucleus can be (i.e. how many neutrons are in the nucleus). We know that the nucleus is composed of neutrons and protons, which we collectively call *nucleons*. An atom of a given element can have a nucleus with a different number of neutrons and have the same chemical behaviour as the other atoms of the same element, to a very good approximation. A nuclide is an atomic species defined by both its atomic number, obtained by counting the number of protons, and the mass number, the total number of protons and neutrons. The atomic number is indicated with Z, and the atomic mass number with A. The atomic number is indicated by the chemical symbol of the element, A is a pre-fixed superscript: for example, the Carbon-14 nuclide is indicated with ^{14}C, indicating that it is made of six protons (hence Carbon) and 14 nucleons in total. $A - Z$ is the number of neutrons. The word *isotope*, from the Greek ἴσος and τόπος, means "same place". Two or more nuclides which have the same Z are called *isotopes* because they occupy the same place in the periodic table. Colloquially, isotopes and nuclides are used as synonyms, but to be correct the term isotope implies a (Table 1.1) comparison between nuclides. The terms which compare nuclides are:

- **isotopes** are nuclides with equal charge Z.
  Example: ^{13}C and ^{14}C have both 6 protons in the nucleus.
- **isobars** are nuclides with same A. The word originates from Greek ἴσος βαρος, same weight
  Example: ^{14}N and ^{14}C have both 14 nucleons, but nitrogen-14 has 7 protons, while Carbon-14 has 6 protons.
- **isotones** are nuclides with the same number of neutrons (A-Z).
  Example: ^{14}C and ^{16}O have both 8 neutrons.

We'll see later that, just like atoms, nuclei can be in excited states, some of which could be meta-stable. Nuclides with the same A and Z, but different excitation states, are called **isomers**.

Isotopes of a given nuclide have the same number of protons and therefore the same number of electrons, as atoms are electrically neutral. So, isotopes have the same chemical properties. Not all isotopes are radioactive, i.e. decay and radiate energy. An element can have two or more stable isotopes. Their chemical properties

Table 1.1 An example of isotope, isobar and isotone nuclides

Isotopes	^{14}C	^{13}C
Isobars	^{14}C	^{14}N
Isotones	^{13}C	^{14}N

Carbon has $Z = 6$ (number of protons), and Nitrogen has $Z = 7$

are mostly the same, but may be slightly different in terms of kinetics of the chemical reactions. For instance, ^{12}C and ^{13}C are both stable isotopes, and they are naturally found in the relative proportion 98.9% and 1.1%. However, their slightly different mass changes slightly the vibrational frequency of the molecules, which makes the chemical reaction speeds to be slightly different for the two isotopes. Living organisms prefer to use ^{12}C rather than ^{13}C, so in living organisms, including their fossil derivatives, like petroleum, we find a slightly lower concentration of ^{13}C, 1.0%. This lower concentration was also found in carbon impurities in apatite rocks in Greenland, which were dated 3.85×10^9 years ago. Thanks to this tiny difference in chemical properties of stable isotopes, we now know that life started 690 million years after the formation of the Earth.

From early investigations, it was soon evident that radioactive materials emit three kinds of radiation: positively charged, (mostly) negatively charged and electrically neutral radiation. They were called α, β and γ (alpha, beta and gamma), respectively (Fig. 1.2). The γ radiation is of the same electromagnetic nature as the infrared, as light, and X-rays, but is emitted by nuclei, while X-rays are emitted by atoms. An X-ray trajectory is not bent by the presence of electric and magnetic fields.

The α particles were soon identified as ^{4}He nuclei, which are made by two neutrons and two protons bound together by the nuclear force. β radiation is made of electrons or their positively charged anti-particle, *positrons* , and are emitted by the transmutation of atomic nuclei. As the γ radiations are electrically neutral, their trajectories are unaffected by the presence of electric or magnetic fields, while the α and β particle trajectories are deflected in opposite directions when immersed in the same magnetic or electric field. By measuring the radius of a charge particle

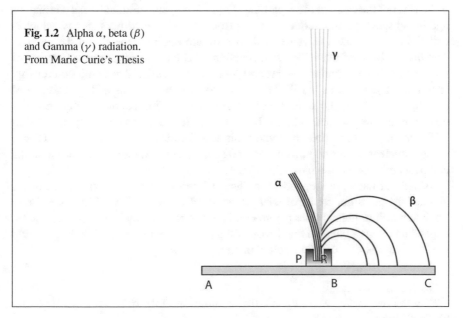

Fig. 1.2 Alpha α, beta (β) and Gamma (γ) radiation. From Marie Curie's Thesis

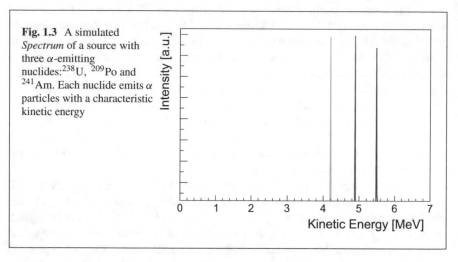

Fig. 1.3 A simulated *Spectrum* of a source with three α-emitting nuclides:^{238}U, ^{209}Po and ^{241}Am. Each nuclide emits α particles with a characteristic kinetic energy

trajectory in a magnetic field, we measure the ratio between the electric charge q and the mass m. Conversely, if we know q/m, we can measure the kinetic energy, and its distribution. In a constant magnetic field, the bending radius is inversely proportional to the momentum of the particle. A distribution of kinetic energies of the emitted particles, for a given source of radiation, is called an *energy spectrum*; a distribution of momenta is a *momentum spectrum*. The α radiation gives rise to spectral "lines", which are characteristic of the emitting nuclide. This means that a given nuclide emits all α particles with a kinetic energy and a momentum in a very narrow range, as shown in Fig. 1.3.

This is not the case for the β particles Fig. 1.4: they are emitted with a continuous and broad spectrum of energies, ranging from zero to a maximum energy, which is called the *end-point*. The value of the maximum energy depends on the nuclide. The distributions of intensities as a function of kinetic energy, or as a function of momentum, have a similar shape for all β-emitting nuclides. The continuous energy spectrum of β particles led W. Pauli[1] to make the hypothesis of the emission of a neutral particle at the same time as the electron. This neutral particle escapes detection and was named *neutrino* by E. Fermi. It was experimentally detected in 1956, about 40 years after the hypothesis was formulated, by Cowan and Reines using a nuclear power plant as a neutrino source. The different behaviour in alpha and beta emission is summarized in Fig. 1.5.

Alpha particles are stopped by a sheet of paper, or a few centimetres of air at standard pressure. Beta particles are stopped by about 2 m of air or a 4-mm thick aluminium plate, while gamma radiation is blocked by lead plates or bricks, depending on their energy (Fig. 1.6). We'll try later to get some introductory insight into the passage of radiation through matter.

---

[1]Wolfgang Pauli, (1900–1958) from Austria. He was awarded the Nobel prize in 1945 for his exclusion principle.

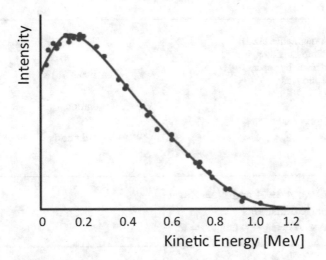

Fig. 1.4 Spectrum of a β source ^{210}Bi, from Neary, M.A. Proc. R. Soc. Lond. A (1940). At the time, ^{210}Bi used to be called Radium-E

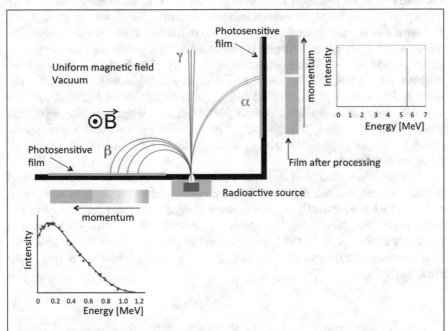

Fig. 1.5 The α particle from a nuclide all have about the same kinetic energy, which is a characteristic of the nuclide. The β particles have a continuous distribution of energies, from zero to a maximum which depends on the nuclide. The momentum spectrum of particles can be measured using a magnetic field which is perpendicular to the plane of the figure. This equipment is called a *spectrometer*

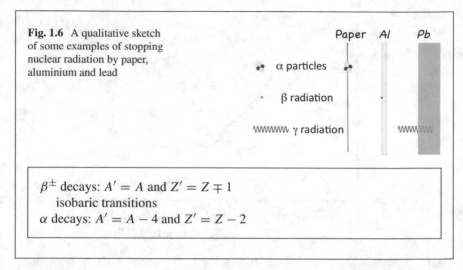

Fig. 1.6 A qualitative sketch of some examples of stopping nuclear radiation by paper, aluminium and lead

$\beta^{\pm}$ decays: $A' = A$ and $Z' = Z \mp 1$
    isobaric transitions
α decays: $A' = A - 4$ and $Z' = Z - 2$

It was found experimentally, soon after their discovery, that beta-emitting decays transform nuclides into their isobars with $Z' = Z \pm 1$, while the emission of an alpha particle transforms the nuclide into another with $A' = A - 4$ and $Z' = Z - 2$.

Like α radiation, γ-radiation, from a given unstable nuclide, is characterised by a narrow energy range of the emitted particle. They are *quanta* of electromagnetic radiation, which are called *photons*, independently of their energy. Typically, γ-emitting decays follow α- or β-emitting decays, which produce *daughter* nuclei in an excited state, which are called *isomers* of the more stable nuclides. Just like atoms, excited nuclei produce electromagnetic radiation when transitioning to their ground state. In general, photons emitted by nuclei or by sub-nuclear particles are called γ rays or γ particles, while photons emitted by atoms or by low-energy electrons are X-rays.

The early classification of α, β and γ radiation still holds after 120 years, with the addition of neutron radiation. Some materials, alone or in a mixture, emit radiation that is electrically neutral, extremely penetrating and very dangerous to living organisms, which was recognised as being due to neutron emission. The neutron was discovered in 1932, quite some time after the proton and the electron. It is the neutral equivalent of the proton. Neutron radiation was discovered in the reaction:

$$\alpha + {}^9\text{Be} \rightarrow {}^{12}\text{C} + n \tag{1.1}$$

where α particles impinge onto a Beryllium target. The α particles originate from a Polonium source, and the neutron source is named after the radioactive nuclide and the target material, Polonium–Beryllium in this case.

Nuclear reactors are intense sources of neutron radiation. Its main effects are nuclear transmutations and activation of materials which were not radioactive before exposure.

At this point, we could start immediately with radioactive decays, the interaction of radiation with matter, a description of the atomic nucleus, the nuclear energy from fission and fusion and some elements of particle physics. However, we do not yet have all the tools to treat the subject. If we use classical physics laws, we can easily get unphysical results. What follows is a quick example.

The radioactive isotope (well, we just agreed to call it nuclide instead) ^{90}Sr produces β particles with a kinetic energy (E) distribution which extends up to 546 keV. Classically, $E = 1/2 m_e v^2$, so the velocity v of the beta particle is $v = \sqrt{2E/m_e}$, where m_e is the mass of the electron, $m_e = 9.109 \times 10^{-31}$ kg.

$$v = \sqrt{\frac{2 \times 546 \times 10^3 \times 1.60 \times 10^{-19}}{9.109 \times 10^{-31}}} \text{m/s} = \sqrt{1.918 \times 10^{17}} = 4.38 \times 10^8 \text{ m/s}.$$

(1.2)

The speed of light in vacuum is $c = 299,792,458 \approx 3.00 \times 10^8$ m/s. Some electrons from that radioactive source would be superluminal, according to classical mechanics. We know that no particle or wave can exceed the speed of light in vacuum. Clearly, we are in the domain of very high velocities and that is why we need to introduce special relativity. Nuclear and subnuclear particles are "small" on an absolute scale, have small masses and move fast.

1.1 Problems

For these problems, rounding to the second decimal digit is more than enough. Their purpose is to make the reader familiar with the system of units which are used in particle and nuclear physics.

α particle mass value: $m_\alpha = 6.4424 \times 10^{-27}$ kg.

Positron mass value: $m_e = 9.109382 \times 10^{-31}$ kg.

Electron electrical charge: $e = 1.6021765 \times 10^{-19}$ C

α particle mass value: $m_\alpha = 3.7273 \times 10^9$ eV/c^2 = 3.7273 GeV/c^2

1.1 An alpha particle (^{4}He nucleus, with charge $q = +2e$) is initially at rest, close to an electrically grounded metal plate. At time $t = 0$, we turn on the electric field between the plate and another metal plate, parallel to it, located at a distance of 0.1 m. They are all in vacuum. The electric potential difference between the plates is -1 kV. What is the kinetic energy of the alpha particle when it reaches the second plate? What is the α particle momentum when it reaches the plate? Compare it with the result obtained when a positron (positive electron) is in the same situation as the alpha particle. Express the momenta in SI units.

1.2 The nuclide ^{241}Am emits α particles with a kinetic energy of 3.5 MeV. Calculate the speed of these ^{4}He nuclei using classical physics formulas; express the result as a ratio to the speed of light ($\beta = v/c$).

1.3 The fruit fly *Drosophila melanogaster* has a mass of about 0.2 mg and can reach a speed of about 0.5 m/s. At this speed, what is its kinetic energy, expressed in eV?

1.4 At the Large Hadron Collider, the particle beam energy is 6.5 TeV. The accelerator stores 2000 *bunches* of particles, each containing 1.1×10^{11} protons. What is the stored energy of a single beam, in Joules.

1.5 A car weighs 1000 kg and runs at 100 km/h. Calculate its kinetic energy in Joule and compare it with the result of problem 1.4.

1.2   Solutions

Solution to 1.1   The kinetic energy is equal to the electric potential difference multiplied by the charge. Therefore, the kinetic energy is

– α particle: $Q_\alpha = 2e$ and its kinetic energy is

$$E_k^{(\alpha)} = Q\Delta V = 2 \text{ keV}.$$

– electron: it has a charge $Q_e = e$ and its kinetic energy is

$$E_k^{(e)} = Q\Delta V = 1 \text{ keV}.$$

To convert these values into SI units, we remember that 1 J = 1 CV. In addition, in classical physics the momentum of a particle of mass m is $p = \sqrt{2mE_k}$.

– For the α particle: $m_\alpha = 6.64 \times 10^{-27}$ kg;

$$E_k = 2 \text{ keV } = 2 \times 10^3 \times 1.60 \times 10^{-19} = 3.20 \times 10^{-16} \text{J}$$

$$p_\alpha = \sqrt{2 \times 3.20 \times 6.64 \times 10^{-27} \times 10^{-16}} = \sqrt{4.25 \times 10^{-42}}$$

$$p_\alpha = 2.06 \times 10^{-21} \text{ kg m/s}$$

– For the positron: its mass is $m_e = 9.11 \times 10^{-31}$ kg;

$$E_k = 1 \text{ keV } = 1.60 \times 10^{-16} \text{ J}$$

$$p_e = \sqrt{2 \times 1.60 \times 9.11 \times 10^{-31} \times 10^{-16}} = \sqrt{2.915 \times 10^{-46}}$$

$$p_e = 1.71 \times 10^{-23} \text{ kg m/s}$$

The α particle and the positron end up with having similar kinetic energy, but very different momenta, due to the large difference in mass. Calculating in SI units requires using large exponents.

Solution to 1.2

$$v = \sqrt{2E_k/m_\alpha} = \sqrt{\frac{2 \times 3.5 \times 10^6 \text{ (eV)}}{3.7273 \times 10^9 \text{ (eV}/c^2)}} = 0.043 \text{ c}$$

The speed of the α particle is 4.3% of the speed of light. This value is high, but classical physics formulas are still a valid approximation, in this case.

Solution to 1.3

$$E_k = 1/2mv^2 = 0.5 * 2 \times 10^{-7} \times 0.5^2 [\text{ kg m}^2\text{s}^{-2}] = 0.25 \times 10^{-7} \text{ J} = 25 \text{ nJ}$$

We know that

$$1 \text{ J} = 1 \text{ CV} = 1/Q_e \text{ eV} = 1/(1.6021765 \times 10^{-19}) \text{ eV}$$

$$E_k = \frac{0.25 \times 10^{-7}}{1.6021765 \times 10^{-19}} = 1.56 \times 10^{11} \text{ eV} = 156 \text{ GeV}$$

A fruit fly has the same energy of a particle accelerated by a large accelerator. However, its kinetic energy is distributed to all its molecules, while in an accelerator it is concentrated in one particle.

Solution to 1.4   Recalling that 1 TeV $= 10^{12}$ eV and 1 eV $= 1.602 \times 10^{-19}$ J, we have:

$$E(\text{beam}) = 6.5 \times 10^{12} \times 2. \times 10^3 \times 1.1 \times 10^{11} \times 1.602 \times 10^{-19} = 2.29 \times 10^8 \text{ J}$$

$$E(\text{beam}) = 229 \text{ MJ}$$

Solution to 1.5   Recalling that 1 km/h $= 1000/3600 = 0.2778$ m/s

$$E_k = 1/2mv^2 = 0.5 \times 10^3 \times (10^2 \times 0.2778)^2 = 3.85 \times 10^5 \text{ J}$$

The car's energy is 385 kJ, about three orders of magnitude less than the energy stored in a single LHC beam. The LHC beam energy is comparable to the kinetic energy of a train.

Chapter 2
Special Relativity

2.1 Introduction

This chapter is included mostly for completeness, because many textbooks already explain special relativity in a deeper and more elegant way Galileo.[1]

The principle of relativity is present in classical physics since its very beginning. Galileo, in his Dialogue Concerning the Two Chief World Systems (1632), states that the laws of physics are the same in all reference frames that are in relative uniform motion. More precisely, Galileo realised that passengers in a ship have no way to tell if the ship is moving or standing.

Shut yourself up with some friend in the main cabin below decks on some large ship, and have with you there some flies, butterflies, and other small flying animals. Have a large bowl of water with some fish in it; hang up a bottle that empties drop by drop into a wide vessel beneath it. With the ship standing still, observe carefully how the little animals fly with equal speed to all sides of the cabin. . . . When you have observed all these things carefully (though doubtless when the ship is standing still everything must happen in this way), have the ship proceed with any speed you like, so long as the motion is uniform and not fluctuating this way and that. You will discover not the least change in all the effects named, nor could you tell from any of them whether the ship was moving or standing still.

---

[1] Galileo Galilei, 1564–1642, is considered the father of modern scientific method. He taught in Pisa and Padua, Italy.

© Springer Nature Switzerland AG 2018
S. D'Auria, *Introduction to Nuclear and Particle Physics*,
Undergraduate Lecture Notes in Physics,
https://doi.org/10.1007/978-3-319-93855-4_2

Rinserratevi con qualche amico nella maggiore stanza che sia sotto coverta
di alcun gran navilio, e quivi fate d'aver mosche, farfalle e simili animaletti
volanti: siavi anco un gran vaso d'acqua, e dentrovi de' pescetti; sospendasi
anco in alto qualche secchiello, che a goccia a goccia vada versando dell'acqua
in un altro vaso di angusta bocca che sia posto a basso; e stando ferma la nave,
osservate diligentemente come quelli animaletti volanti con pari velocità vanno
verso tutte le parti della stanza. [..][Osservate che avrete diligentemente tutte
queste cose, benché niun dubbio ci sia mentre il vascello sta fermo non debbano
succedere così: fate muovere la nave con quanta si voglia velocità; ché (pur di
moto uniforme e non fluttuante in qua e in là) voi non riconoscerete una minima
mutazione in tutti li nominati effetti; né da alcuno di quelli potrete comprendere
se la nave cammina, o pure sta ferma.

What is new in Einstein's relativity is that, in order to make electrodynamics
invariant in passing from one inertial frame to another, an additional ingredient is
required: that there is one special speed, which is the speed of light in vacuum. This
speed is required to be exactly the same in all inertial frames and cannot be surpassed
by any particle carrying energy. This hypothesis has been proved experimentally by
a famous experiment by Michelson and Morley. The Earth orbits around the sun
at a variable speed of about 30 km/s with respect to the sun. This is still 0.01%
of the speed of light in vacuum. In addition, the solar system orbits in one arm
of our spiral galaxy, the Milky Way, at about 220 km/s with respect to the black
hole at the centre of our galaxy. We also know that the galaxies next to ours are
moving towards the "Great Attractor", a structure in the intergalactic space, at about
1000 km/s. This is 0.3% of the speed of light. If the speed of light changed in our
reference frame on the Earth, we would notice an effect which changes daily in
our laboratory, due to the Earth's rotation around its axis. Now, an effect of 0.01%
may seem extremely difficult to measure directly. The light covers distance of 3 km
in just 10 μs, in vacuum, so an effect of 0.01% corresponds to a time of 1 ns,
which is well within the range of modern measurements of time. The experiment
could be repeated nowadays with a laser as a direct measurement, but this is not
what Michelson and Morley did, back in 1887. Rather they measured a distance,
on an optical bench. In terms of distance, on an optical path of only 1 m 0.01%
is 10^{-4} m, or 0.1 mm. Compared to the wavelengths of Sodium light, 589.0 and
589.6 nm, this distance is about a 1000 times larger. Thanks to the short wavelength
of visible light, it was possible to measure precisely that there was no variation
of the speed of light in the 24-h period, or even when rotating the optical bench,
which was floating on a mercury bath (this would not be allowed today). Hendrik

Fig. 2.1 The map of the Cosmic Microwave Background Radiation (NASA)

Lorentz[2] had already found and published[3] the mathematical formulas to allow the speed of light to be constant; Henri Poincaré refined the formulae, which he called "Lorentz transformations", but it was Albert Einstein, in 1905, who clarified the subject and gave the exact explanation in his article "On the electrodynamics of moving bodies".[4] The two principles of special relativity are:

1. The laws of physics are the same in all inertial reference frames. There is no inertial reference frame which is better than others to describe the laws of physics.
2. The speed of light is the same in all reference frames.

We'll see that there are other effects which allow us to measure relative speeds: relative motion does not change the speed of light but changes its frequency, which corresponds to its colour, as will be shown later. This is known as red or blue shift effect: specific feature of light from distant stars appears to us at a different frequency owing to the relative speed. Nowadays, we can find one reference frame which is somehow special: the frame where the Cosmic Microwave Background Radiation (CMBR (Fig. 2.1)) is uniform and isotropic, to a first approximation; however, it is not better than others to describe the laws of physics.

---

[2]Hendrik Lorentz, 1853–1928, was professor in Leiden, the Netherlands. He was awarded the Nobel prize in 1902 for the explanation of the Zeeman effect.

Henry Poincaré, France (1854–1912) was a mathematician and theoretical physicist.

[3]H.A. Lorentz, Simplified Theory of Electrical and Optical Phenomena in Moving Systems, in: KNAW, Proceedings, 1, 1898–1899, Amsterdam, 1899, pp. 427–442;

"Electromagnetic phenomena in a system moving with any velocity smaller than that of light", Proceedings of the Royal Netherlands Academy of Arts and Sciences, 6: 809–831 (1904).

[4]"Zur Elektrodynamik bewegter Körper", Annalen der Physik 17: 891 (1905).

By measuring the frequency shift in opposite directions, we can measure the velocity of the Earth with respect to this background radiation. The result is that the Earth is moving at about 390 km/s towards the Leo constellation, which is located not far from Ursa Major. If you are in the northern hemisphere, you can start from the Ursa's two stars which form the outer bowl of the Big Dipper, or Plough, and draw a straight line; to the north, you'll find Polaris, while Leo is about the same distance from Ursa Major, but at the opposite side.

2.2   The Lorentz Transformations

We'll now derive the correct way to change variables from one inertial reference frame S to another inertial reference frame S', which is moving at constant velocity $\vec{u}$ with respect to S, based on the principles of special relativity. We assume that each reference frame uses Cartesian coordinates, which for the system S are (x, y, z); the time is measured in each frame, with clocks which are at rest in that frame. In the frame S, we'll measure time t. The reference frame S' has coordinates (x', y', z', t'), where with t' (Fig. 2.2a) we mean that the time is measured with clocks which are at rest in the frame S' (Fig. 2.2a). The clocks are synchronised between the two frames: assuming that normal synchronisation occurs within each reference frame, synchronisation occurs between clocks at the same spatial position: say, a synchronisation signal is exchanged between the two clocks at the origin at the time when the origins of the two reference frames are in the same position.

The rules to correctly transform the mathematical description of a physics "event" in the frame S to a description in the frame S' are called the Lorentz transformations, after the Dutch physicist Hendrik Antoon Lorentz, who wrote these transformations well before Einstein correctly explained them.

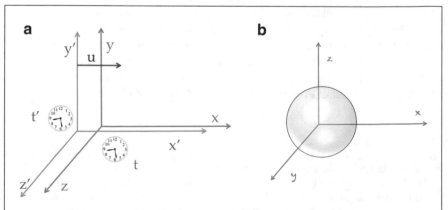

Fig. 2.2 (**a**) Reference frame S' is moving with constant velocity $\vec{u}$ with respect to the reference frame S. Each reference frame has its own synchronised clocks. (**b**) The wavefront of a light pulse must be a sphere in all reference frames

Let's state some desirable features of these transformations. They must only depend on the relative velocity of the two frames $\vec{u}$. When this velocity is small, compared to the speed of light, c, the Lorentz transformations must be approximated by the Galileo–Newton transformations. Without lack of generality, we can rotate the reference frames S and S' in such a way that $\vec{u}$, $\hat{x}$ and $\hat{x}'$ are parallel to each other. The Galileo–Newton transformations are

$$x' = x - ut \tag{2.1}$$

$$t' = t \tag{2.2}$$

The Lorentz transformations must be linear, otherwise they would distort the space–time. We cannot use higher powers of x, t, or other functions, like exponentials or logarithms. A straight line in one frame must be a straight line in the other frame. A uniform motion in one reference frame must be a uniform motion in the other reference frame. So, the transformations must depend on x, y, z, t, but not on any product like xy or x^2. Also, for $\vec{u} \to 0$, the transformation must be the identity transformation. The transformations must be symmetrical by exchange $x \leftrightarrow x', y \leftrightarrow y', z \leftrightarrow z', \vec{u} \leftrightarrow -\vec{u}$: the two reference systems are interchangeable.

It is reasonable to expect that the directions perpendicular to the relative motion are unaffected, just like in Galileo transformations: $y' = y$ and $z' = z$. Combining all requirements from above, we can write a generic transformation for space coordinate:

$$x' = \gamma(x - ut) \tag{2.3}$$

and

$$x = \gamma(x' + ut) \tag{2.4}$$

where we need to determine the factor γ. If we require that the speed of light is the same in both reference frames (we'll call it c), the time coordinate cannot be the same in the two frames: it must transform. The same arguments as above are applied to the time coordinate: we can write a generic linear transformation which reads as $t' = b_1 x + b_2 t$. We have three parameters to determine: γ, b_1, b_2. All together, we have

$$t' = b_1 x + b_2 t \tag{2.5}$$

$$x' = \gamma(x - ut) \tag{2.6}$$

$$y' = y \tag{2.7}$$

$$z' = z \tag{2.8}$$

From the second postulate, we must now require that a light pulse that radiates in all directions should be described exactly in the same way in both frames. Let's assume that at a given time $t = t' = 0$ the two reference frames overlap, and a light bulb is turned on at the origin of both frames. The wavefront of the light must be described by a sphere in both reference frames (Fig. 2.2b): observing from the frame S, we have

$$x^2 + y^2 + z^2 = c^2 t^2 \tag{2.9}$$

while in the frame S':

$$x'^2 + y'^2 + z'^2 = c^2 t'^2 \tag{2.10}$$

We substitute (2.5)–(2.8) into (2.10), and we get

$$\gamma^2 (x - ut)^2 + y^2 + z^2 = c^2 (b_1 x + b_2 t)^2 \tag{2.11}$$

which becomes

$$(\gamma^2 - c^2 b_1^2) x^2 + y^2 + z^2 = (c^2 b_2^2 - \gamma^2 u^2) t^2 + (2\gamma^2 u + 2c^2 b_1 b_2) tx \tag{2.12}$$

The formula above is the equation of a sphere with radius ct only if:

$$2\gamma^2 u + 2c^2 b_1 b_2 = 0; \tag{2.13}$$

$$\gamma^2 - c^2 b_1^2 = 1; \tag{2.14}$$

$$c^2 b_2^2 - \gamma^2 u^2 = c^2 \tag{2.15}$$

Rearranging the last two equations:

$$b_1^2 = \frac{\gamma^2 - 1}{c^2} \tag{2.16}$$

$$b_2^2 = 1 + \gamma^2 \frac{u^2}{c^2} \tag{2.17}$$

and rearranging and squaring Eq. (2.13), we have

$$\gamma^4 u^2 = c^4 b_1^2 b_2^2 \tag{2.18}$$

Substituting the values of b_1^2 and b_2^2, we obtain

$$\gamma^4 u^2 - c^4 \frac{\gamma^2 - 1}{c^2} \left(1 + \gamma^2 \frac{u^2}{c^2} \right) = 0; \tag{2.19}$$

From (2.19), we derive

$$\gamma^4 u^2 - c^2 \gamma^2 - \gamma^4 u^2 + c^2 + \gamma^2 u^2 = 0$$

$$\gamma^2 (u^2 - c^2) + c^2 = 0$$

$$\gamma^2 = \frac{c^2}{c^2 - u^2} = \frac{1}{1 - \frac{u^2}{c^2}}$$

$$\gamma = \frac{1}{\sqrt{1 - \frac{u^2}{c^2}}} \tag{2.20}$$

Substituting into (2.17), we have $b_2^2 = \gamma^2$, and we must choose $b_2 = +\gamma$ for continuity. This means that for very small values of u the Lorentz transformations must be well approximated by the Galileo–Newton transformations. From (2.16), we have

$$b_1^2 = \frac{\gamma^2 - 1}{c^2} = \frac{u^2}{c^4} \left(\frac{1}{1 - \frac{u^2}{c^2}} \right) \tag{2.21}$$

This time (for the same reason as above), we have to choose $b_1 = -\frac{u}{c^2}\gamma$ so that the Lorentz transformations are

$$x' = \gamma(x - ut); \tag{2.22}$$

$$y' = y; \tag{2.23}$$

$$z' = z; \tag{2.24}$$

$$t' = \gamma\left(t - \frac{u}{c^2}x\right) \tag{2.25}$$

where

$$\gamma = \frac{1}{\sqrt{1 - \frac{u^2}{c^2}}} \tag{2.26}$$

It is evident from the transformations that in the case where the velocity $u \ll c$ we have $\gamma \approx 1$, $u/c^2 \approx 0$ and we find the Galileo–Newton transformations. This allows us to keep the speed of light in vacuum constant in all reference frames. The speed of light is nowadays defined to be exactly 299792458 m/s. The price to pay to have a constant value for c in all reference frames is that time is not the same in all reference frames. Events which are seen as contemporary in one reference frame are no longer happening at the same time in another frame. However, what is preserved is causality: if one event is causing another, there is no reference frame where it may occur after its effect.

Suppose we turn on a light bulb in the middle of a train carriage at local time $t_0 = 0$. In the reference frame of the carriage where the bulb is at rest, the light will reach both ends of the carriage at the same time $t_1 = l/(2c)$, where l is the length of the carriage. If the carriage is moving at speed $\vec{u}$ with respect to the station, the light pulse will be seen to reach the rear end at time $t_1' = (l - u\Delta t')/c$ and the forward end of the carriage at time $t_2' = (l + u\Delta t')/c$. Actually, with this example,

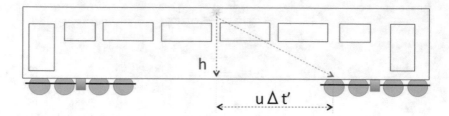

Fig. 2.3 Time intervals are different when observed in reference frames which are in motion one with respect to the other. This figure illustrates the example of the light bulb in a train carriage

we can derive the time contraction formula: suppose the bulb is on the ceiling of the carriage, at a height h from the floor. When observed within the carriage, the time needed by the light to reach the floor is simply $\Delta t = h/c$.

When observed from the station, while the light is in flight, the floor has moved by a distance $d = u\Delta t'$ (Fig. 2.3), so that the light had to go on a straight line, diagonal this time, with a length $h' = \sqrt{h^2 + u^2\Delta t'^2}$. The time to reach the floor, as seen from the station, is $\Delta t' = h'/c$. By squaring, we have

$$\Delta t'^2 = \frac{1}{c^2}(h^2 + u^2\Delta t'^2);$$

using $h^2 = c^2\Delta t^2$, we have

$$\Delta t' = \frac{\Delta t}{\sqrt{1 - \frac{u^2}{c^2}}} = \gamma\Delta t; \tag{2.27}$$

The formula above is very important for practical uses: it describes the time dilation when observing a phenomenon in a reference frame where the source of the phenomenon is not at rest.

As $\gamma \geq 1$, times are dilated: moving clocks are running slower than clocks at rest. In this case, the moving clock is the one in the station: we consider at rest the clock which is in the frame where the light bulb is at rest. The relativistic γ factor is very close to one for velocities up to about 30–40% of the speed of light, (Fig. 2.4) and this explains why we do not observe relativistic effects in everyday life. However, nuclear particles have speed comparable to the speed of light and special relativity must be used. The relativity formulas are verified in thousands of applications, including those involving particle accelerators. It is sometime useful to picture a motion in a space-time graphics, which is also called a Minkowski diagram, as shown in Fig. 2.5.

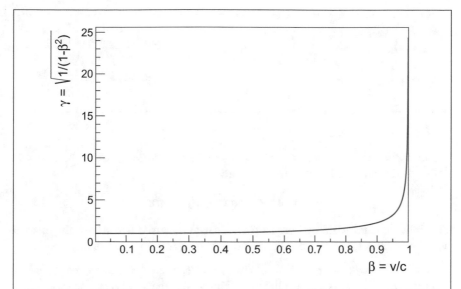

Fig. 2.4 The relativistic γ factor is very close to one for velocities up to about 30–40% of the speed of light, but increases rapidly above $\beta = 0.9$; this plot is truncated, it is clear from its definition that $\gamma \to \infty$ for $\beta \to 1$

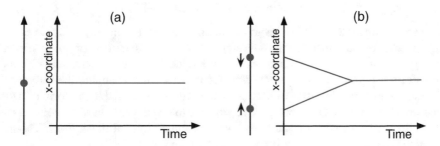

Fig. 2.5 (**a**) A particle at rest in the space–time plot. (**b**) A graphical representation in the space–time plane of two particles moving in opposite direction on the same straight line and then remaining at rest as one particle, in a completely inelastic collision

2.3 Velocity, Momentum and Energy, 4-Vectors

We have seen that the time duration of a phenomenon is larger when observed from a moving reference frame, with respect to the reference frame where the phenomenon (or its origin) is at rest. Let's consider two events recorded in the reference frame S: one with coordinate (t_1, x_1, y_1, z_1) and another with coordinates (t_2, x_2, y_2, z_2). We can define the space–time interval as:

$$\Delta s^2 = c^2 \Delta t^2 - \Delta x^2 - \Delta y^2 - \Delta z^2 \tag{2.28}$$

	ds^2 value	Interval
Table 2.1 Naming of the intervals in space–time	$ds^2 = 0$	Light-like
	$ds^2 > 0$	Time-like
	$ds^2 < 0$	Space-like

where $\Delta x = x_2 - x_1$ etc. For two events extremely close to each other, we can define the infinitesimal space–time interval:

$$ds^2 = c^2 dt^2 - dx^2 - dy^2 - dz^2 \qquad (2.29)$$

This interval is invariant by Lorentz transformations: given two events, the interval between them is the same no matter what reference frame is used to measure it.

If $c^2 dt^2 > dx^2 + dy^2 + dz^2$, the event separation is said to be *light-like* (Table 2.1). There is an ideal light ray which joins the two events.
If $c^2 dt^2 > dx^2 + dy^2 + dz^2$ (in this definition $ds^2 > 0$), the space–time interval is *time-like*.
If $c^2 dt^2 < dx^2 + dy^2 + dz^2$ (in this definition $ds^2 < 0$), the space–time interval is *space-like*. No cause–effect relationship can exist between the two events, they can happen at the same time in some reference frame.

The space–time interval is left invariant by Lorentz transformations, it is "Lorentz-invariant". We can introduce the 4-vectors as sets of 4 quantities, or components, of which 3 are space-related and one is time-related. These mathematical objects transform according to the Lorentz transformations when changing the reference frame. As an example space–time coordinates can be expressed in the form of a 4-vector, which is indicated in boldface $\mathbf{x}$. We have to make sure to have homogeneous quantities: all four components must have the same physical dimensions. This is obtained by multiplying the time component by the speed of light, c.

$$\mathbf{x} = (ct, \vec{x}) = (ct, x, y, z) \qquad (2.30)$$

Let's look at velocities now, to see if they form a 4-vector. For a particle of mass m:

$$v_x = \frac{dx}{dt};$$

$$v'_x = \frac{dx'}{dt'} = \frac{\gamma(dx - udt)}{\gamma(dt - \frac{v}{c^2}dx)}) = \frac{dx/dt - u}{1 - \frac{u}{c^2}\frac{dx}{dt}}$$

$$v'_x = \frac{v_x - u}{1 - \frac{u}{c^2}v_x} \qquad (2.31)$$

The other components of the velocity are also affected by the Lorentz transformations, via the dt term:

$$v'_y = \frac{v_y}{\gamma(1 - \frac{u}{c^2}v_x)} \qquad\qquad v'_z = \frac{v_z}{\gamma(1 - \frac{u}{c^2}v_x)} \qquad (2.32)$$

The velocity (as defined above) does not transform according to the Lorentz transformations and is not part of a 4-vector. However, we do have a 4-vector if we define the velocity as a derivative with respect to the *proper time* τ $d\tau = \frac{1}{\gamma}dt$. The proper time is the time as measured in the reference frame where the particle is at rest. In this case, the 4-velocity has simply c as the "time" component. The momentum is more interesting. We define the relativistic 3-momentum ($\vec{p}$) of a particle as:

$$\vec{p} = m\frac{d\vec{x}}{d\tau} = \gamma_v m\vec{v} \qquad (2.33)$$

The γ factor in the above formula is one of the reference frames where the particle is at rest. We can add a time-like component, which we call relativistic energy $p_0 = E = \gamma mc^2$ and show that $(p_0, \vec{p}) = (E/c, p_x, p_y, p_z)$ transforms in the same way as the space and time coordinates, i.e. is a 4-vector. We call it 4-momentum. 4-vectors can be added, can be multiplied by a numeric factor and we can define a scalar product:

$$(x_0, x_1, x_2, x_3)(y_0, y_1, y_2, y_3) = x_0 y_0 - x_1 y_1 - x_2 y_2 - x_3 y_3. \qquad (2.34)$$

The scalar product between 4-vectors can be considered as a standard scalar product when we multiply one of the two vectors by a matrix g:

$$g = \begin{pmatrix} 1 & 0 & 0 & 0 \\ 0 & -1 & 0 & 0 \\ 0 & 0 & -1 & 0 \\ 0 & 0 & 0 & -1 \end{pmatrix}$$

which is called metric matrix and changes sign to the space coordinates.

The scalar products of 4-vectors are relativistic invariants, so we can define a relativistically invariant 4-vector modulus. In particular, the modulus of the 4-momentum is a relativistic invariant, and its value is left invariant by any process, like a radioactive decay. The modulus of the 4-momentum of the parent nucleus/particle is the same as the modulus of the sum of the 4-momenta of the decay products. This is the *invariant mass* of a system of particles. Let's consider a scattering process $A + B \rightarrow C + D + E$ or a α decay $A \rightarrow B + \alpha$, where A, B, C, D, E are particles or nuclei. We are used to 3-momentum conservation, where the sum of the momentum of the right-hand side is equal to the sum of the momentum of the left-hand side.

$$\vec{p_A} + \vec{p_B} = \vec{p_C} + \vec{p_D} + \vec{p_E} \tag{2.35}$$

and

$$\vec{p_A} = \vec{p_B} + \vec{p_\alpha} \tag{2.36}$$

We write the conservation of the 4-vector energy–momentum in exactly the same way:

$$\mathbf{p}_A + \mathbf{p}_B = \mathbf{p}_C + \mathbf{p}_D + \mathbf{p}_E \tag{2.37}$$

and for a decay

$$\mathbf{p}_A = \mathbf{p}_B + \mathbf{p}_\alpha \tag{2.38}$$

This is a compact way to state the energy and momentum conservation. The equation above is valid in all inertial reference frames, provided that we transform the 4-momenta according to the Lorentz transformations. The modulus of the sum of 4-vectors has to be the same in both sides of the reaction, and it is the same in all inertial reference frames, so we say it is *invariant*.

$$|\mathbf{p}_B + \mathbf{p}_\alpha|^2 = |\mathbf{p}_A|^2 = m_A^2 c^2 \tag{2.39}$$

From Eq. (2.39) above, we see that this modulus, in case of a decay, is the square of the mass of the initial particle, and therefore it is called *invariant mass*. In particle physics, a method to detect short-lived particles, like the Z^0 or the Higgs boson, is to calculate the invariant mass of the decay products: the obtained values cluster around the mass of the particle which decays (see Fig. 6.10). In case of a decay of a particle "A" into N particles, the sum has to be extended to all the decay products:

$$A \rightarrow \text{particle}_1 + \cdots + \text{particle}_N \tag{2.40}$$

$$|(E_A/c, p_{x_A}, p_{y_A}, p_{z_A})|^2 = \left| \sum_{i=1}^{N} (E_i/c, p_{x_i}, p_{y_i}, p_{z_i}) \right|^2 = m_A^2 c^2. \tag{2.41}$$

2.4 Relativistic Energy

We have defined the relativistic energy of a particle of mass m moving at velocity $\vec{v}$ as $p_0 \equiv E \equiv \gamma_v m c^2$. We need to show that it has something to do with the standard definition of energy in classical physics. Taking the limit $v \ll c$:

$$E = mc^2\gamma = mc^2(1 - v^2/c^2)^{-1/2} \tag{2.42}$$

$$\approx mc^2\left(1 + \frac{1}{2}\frac{v^2}{c^2}\right) \tag{2.43}$$

$$= mc^2 + \frac{1}{2}mv^2 \tag{2.44}$$

we find that it correctly corresponds to the particle kinetic energy plus a constant term, which is fine because only energy differences are important. In the case when $\vec{v} = 0$, we have that the energy at rest of a particle is its mass:

$$E = mc^2 \tag{2.45}$$

This quantity is also the modulus of the energy–momentum 4-vector $\mathbf{p} = (E/c, \vec{p})$, and therefore it is a relativistic invariant. It is the *invariant mass*. From the equation above, we can measure the mass in terms of energy. A process that normally occurs in high-energy physics experiments is the transformation of kinetic energy of particles into mass of new particles and their kinetic energy. This process also occurs naturally when cosmic rays hit the atmosphere. Another example is given by high-energy photons: they have no mass, but they carry energy. In the electric field of a nucleus, they can convert into a pair of charged particles, which have mass. We'll see later the fusion and fission processes, which convert the mass of atomic nuclei into kinetic energy, and ultimately thermal energy.

The masses of subnuclear particles are typically measured in MeV/c^2 or in GeV/c^2. A self-consistent unit system, which is used in theoretical physics, chooses $c = 1$, a dimensionless constant (and also $\hbar = 1$, but this constant will be introduced later). In many text books, this convention is used, and both masses and momenta are measured in units of eV.

It is also evident that the relativistic 3-momentum $\vec{p} = \gamma m\vec{v} \rightarrow m\vec{v}$ becomes the non-relativistic 3-momentum when $v \ll c$ and $\gamma_v \approx 1$.

For free particles, it is fair to assume that the energy and momentum are conserved. Suppose now that we also deal with forces which conserve energy and momentum. Then, from the conservation of our 4-momentum we can show that our new definition of energy is really what we need in terms of Newton's law in some of its various forms.

$$\frac{d}{dt}\mathbf{pp} = 0 = \frac{d}{dt}\left(\frac{E^2}{c^2} - \vec{p} \cdot \vec{p}\right) = \tag{2.46}$$

$$= \frac{2E}{c^2}\frac{dE}{dt} - 2\vec{p} \cdot \frac{d\vec{p}}{dt} = \tag{2.47}$$

$$= 2m\gamma \frac{dE}{dt} - 2m\gamma \vec{v} \cdot \frac{d\vec{p}}{dt} \text{ because } E = m\gamma c^2 \text{ and } \vec{p} = \gamma m\vec{v}$$

$$\tag{2.48}$$

$$= 2m\gamma \left(\frac{dE}{dt} - \vec{v}\frac{d\vec{p}}{dt} \right) \tag{2.49}$$

So, we have demonstrated that for conservative forces, with our relativistic definition of energy we indeed have

$$\frac{dE}{dt} = \vec{v} \cdot \frac{d\vec{p}}{dt} \tag{2.50}$$

The relativistic *kinetic energy* for our free particle is obtained from the relativistic total energy by subtracting the energy of the particle at rest:

$$E_k = \gamma_v mc^2 - mc^2 = (\gamma_v - 1)mc^2 \tag{2.51}$$

The modulus of the energy–momentum 4-vector is Lorentz invariant. Calculating it in the particle rest frame, it is

$$E^2 - \vec{p}\vec{p}c^2 = m^2c^4 \tag{2.52}$$

For massless particles (i.e. the *photon*, the only[5] massless particle we know of), $E^2 = p^2c^2$. It will be shown later that the energy of a photon is given by its frequency: $E_\gamma = h\nu$. The photon carries a momentum which is $p_\gamma = \frac{h\nu}{c}$. Massless particles in vacuum can only travel at the speed of light. While the *photon* is the only particle with exactly zero mass, other particles, the *neutrinos*, are massless with a good approximation. Neutrinos are produced in beta decays, we do not know exactly their mass, but we know it is very low.

2.5   Doppler Effect

At this point, we need to know what happens to photons when observed from different reference frames. An electromagnetic plane wave can be described by:

$$A(\vec{x}, t) = A_0 \sin (\vec{k}\vec{x} - \omega t) . \tag{2.53}$$

---

[5]The other particle which we believe to be massless is the *gluon*, the messenger of the strong force, which is confined within *mesons* and *baryons* and cannot travel freely, because of the "colour confinement".

We rename $k_0 = \omega/c$ the quantity $(k_0, \vec{k})$ must be a 4-vector if we require that the phase must be the same in all moving frames: $\vec{k}\vec{x} - k_0 ct = \vec{k}'\vec{x}' - k_0'ct'$. This is a *light-like* 4-vector. Suppose we have a plane wave along the x-axis, the same direction as the relative velocity of the two reference frames $\vec{u}$.

$$k'_x = \gamma_u \left(k_x \pm \frac{u}{c} k_0 \right) ; \text{ we can call } \beta = u/c, k = k_0 = 2\pi c \nu \tag{2.54}$$

$$\nu' = \gamma \nu (1 \pm \beta) = \nu \frac{1 \pm \beta}{\sqrt{1 - \beta^2}} = \nu \sqrt{\frac{1 \pm \beta}{1 \mp \beta}} \tag{2.55}$$

When we observe the light from a star, we can measure its speed relative to the Earth by measuring the frequency of characteristic emission lines. In general, this is shifted towards lower frequencies, indicating that stars are moving away from us. As the red colour is located at the lower end of the colour frequency spectrum, this is called *red shift*. The relativistic Doppler effect is completely symmetrical if we exchange source and observer, as it should be. In the acoustic Doppler effect, the motion is relative to the medium, so it does make a difference whether the source is moving or the observer (listener) is moving (Fig. 2.6(a)).

In addition to the longitudinal Doppler effect, there is also the relativistic transverse Doppler effect: the frequency changes also when the observer is moving parallel to the optical wavefront (Fig. 2.6(b)).

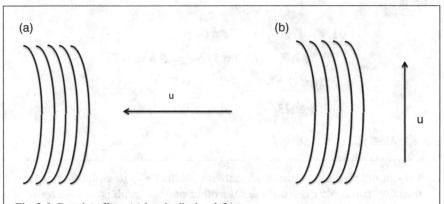

Fig. 2.6 Doppler effect: (**a**) longitudinal and (**b**) transverse

2.6   Group Theory in a Nutshell

The 4-vectors are mathematical objects "living" in a vector space, which is called
Minkowski space.[6] We can add them, multiply by a scalar number and/or make
scalar numbers out of them. Then, we have several *mappings* of 4-vectors onto 4-
vectors: we can rotate them in space, with normal rotations, we can translate in
space and in time, we can "boost" them with velocity $\vec{u}$, i.e. we can observe them
from a reference frame with velocity $\vec{u}$ with respect to the previous one. In general,
all these operations on 4-vectors can be parametrised by one or more parameters:
rotations are parametrised by the angle θ or ϕ, Lorentz boosts by a velocity $\vec{u}$ and
so on. We indicate with $T(m)$ such generic transformations. The sets $\mathcal{G}$ of these
transformations become some mathematical objects, and they can start living their
own mathematical life, independently of the vector space where we initially defined
them. It is reasonable to require that all these transformation must be invertible,
meaning that we can always return to the initial coordinate system; that we can
combine these transformation one after the other and still obtain a transformation;
and that when combining several transformations, we can associate two or more
of them, while keeping the same order. The fact that exists a transformation that
leaves the system unchanged is almost trivial, but required. If the above conditions
are satisfied, the set of transformations is said to form a *group*. More formally:

Even more formally, let $\mathcal{G}$ be a set of transformations and $\circ$ a *composition*
operation

$$1) \forall S, T \in \mathcal{G}, (S \circ T) \in \mathcal{G}$$

$$2) \forall R, S, T \in \mathcal{G}, (R \circ S) \circ T = R \circ (S \circ T)$$

$$3) \exists I \in \mathcal{G} \ni \forall T \in \mathcal{G}, I \circ T = T$$

$$4) \forall T \in \mathcal{G} \exists T^{-1} \ni T \circ T^{-1} = I$$

we say that $(\mathcal{G}; \circ)$ is a *group*.

---

1) We can define a *composition* of transformations ("$\circ$") to combine any two of
    them: we can apply them one after the other, and the result is still the result of a
    transformation of the set;
2) The composition of transformations is associative;
3) There is an identity transformation, which, when composed, with any of the
    others, leaves them invariant;
4) For all transformations, there is an inverse transformation, which is still part of
    the set;

---

[6]After the mathematician Hermann Minkowski, 1864–1909. He taught at Königsberg, Bonn,
Göttingen and Zürich, where Einstein was one of his students.

If, in addition, the composition of transformations commute, i.e. $S \circ T = T \circ S$, the group is said to be *commutative* or *Abelian*.[7] If the transformations, which are the elements of the group, depend on some continuous parameters, for instance the rotation angles, the group is called a *Lie group*.[8] Transformations can be represented by matrices. In case of Lorentz transformations, we have 4×4 matrices of $\mathbb{R}$eal numbers. Rotations in the three-dimensional space are represented by 3×3 matrices.

Lorentz transformations for boosts along the x coordinate have the following matrix form:

$$b_x = \begin{pmatrix} \gamma & -\beta\gamma & 0 & 0 \\ -\beta\gamma & \gamma & 0 & 0 \\ 0 & 0 & 1 & 0 \\ 0 & 0 & 0 & 1 \end{pmatrix},$$

where $\beta = u/c$.

The set made of the Lorentz transformations and the rotations form a group, which is called the *Lorentz group* $\mathcal{L}$. While rotations form a subgroup of $\mathcal{L}$, generic Lorentz transformations don't. However, Lorentz transformations along each of the three axes do form, each in its own, a subgroup of $\mathcal{L}$. A mathematical digression: given a square matrix M its transposed matrix is obtained by swapping its rows with its columns: $M = (M_{ij})$, $M^T = M_{ji}$. A matrix O is called *orthogonal* if $O^T O = \mathbb{1}$ is the diagonal unit matrix. A matrix is said to be *special* if its determinant is equal to 1. The orthogonal matrices in three dimensions with determinant 1 represent the rotation group, which is called SO(3). The Lorentz group is also indicated as $SO^+(1, 3)$. Rotations depend on three parameters, and Lorentz transformations also depend on three parameters, so an element of the Lorentz group is specified by six real parameters.

Analogously to real matrices, complex matrices $N \times N$ can be transposed and conjugated $(M^\dagger)_{ij} = \bar{M}_{ji}$. Here, ‾indicates complex conjugation: $z = (a+ib)$; $\bar{z} = (a - ib)$. A matrix U is said to be *unitary* if $UU^\dagger = U^\dagger U = \mathbb{1}$. If, in addition, the determinant is equal to 1, the group they form is called SU(N), indicating special unitary matrices $N \times N$. Groups can be defined in an abstract way, independently of the vector space and the transformations where we started from. They can act on several *vector spaces*. Given a vector space, a group $\mathcal{G}$ has a certain *representation* in that space: a set of invertible matrices *represent* all the transformations of that group.

A representation of a transformation is *completely reducible* if all matrices are the form of block matrices, which are zero in the off-diagonal blocks. In this case,

---

[7]After Niels Henrik Abel, 1802–1829 Norwegian mathematician. Also, the Abel prize for mathematics is named after him.

[8]After M. Sophus Lie, (pron. as Lee) 1842–1899, Norwegian mathematician, who taught in Oslo and Leipzig.

the vector space, originally of dimension N_0, is divided into invariant subspaces each of dimension N_1 and N_2, with $N_0 = N_1 + N_2$. The representation $\mathbf{R}_0$ is said to be the *direct sum* of two representations $\mathbf{R}_1$ and $\mathbf{R}_2$, and we write

$$\mathbf{R}_0 = \mathbf{R}_1 \oplus \mathbf{R}_2 \quad \text{e.g. } \mathbf{4} = \mathbf{3} \oplus \mathbf{1} \tag{2.56}$$

where the boldface number indicate the dimension of the representation. The number of parameters is still the same as in the original group which is represented. We can form the *direct product* of two or more groups, which is a group depending on a number of parameters given by the sum of the parameters of each group. A representation of this group is also a block matrix, in which each block is a representation of the corresponding sub-group. We write in this case

$$\mathbf{R} = \mathbf{R}_1 \otimes \mathbf{R}_2 \tag{2.57}$$

We have introduced an empty space, some free particles at rest or in uniform motion and mathematical objects like 4-vectors and their transformations, which form groups. In these ideal conditions, many quantities are conserved. The interesting part comes when we introduce the fundamental *interactions*.

2.7   Symmetries and Conserved Quantities

In 1915, Emmy Nöther (Fig. 2.7) proved a fundamental theorem. In general, a physical system with interacting particles can be described by equations in the Lagrangian or Hamiltonian formulation. It is beyond the scope of this book to introduce these, but some of the readers may be familiar with them. The important issue is that given a system, its Lagrangian (or Hamiltonian) equation fully describes its evolution. If these equations don't change when we perform any of the coordinate transformations which are elements of a continuous group with N parameters, then our physical system has N independent conserved quantities. These are also called *"first integrals"*.

This theorem applies equally well to spining tops and elementary particles. Its derivation is beyond the scope of this course, but beautiful proofs of it can be found in several textbooks. Some notable examples are: if the Lagrangian of our system is independent of time, then the energy is conserved. If the Lagrangian equation remains invariant under translations in space, then the momentum is conserved; if it does not depend on the particular orientation, then the angular momentum is conserved. This theorem is very important and has inspired the theories of fundamental interactions: the interactions can be introduced starting from the relativistic quantum mechanical Lagrangian equation of non-interacting particles and requiring its invariance with respect to groups of local transformations. A transformation is local when the parameters of the transformation are a smooth

Fig. 2.7 Amalie Emmy Nöther in 1930, in Koenigsberg. She was born in Germany in 1885 to a Jewish family, and she taught in Goettingen and for a short period in Moscow. In 1933, she had to move to the USA to continue her activity as a mathematician. She prematurely died 2 years later

An example, a representation of the group SO(2) in four dimensions, which is a direct sum has the following matrix form:

$$R(\theta) = \begin{pmatrix} \cos\theta & -\sin\theta & 0 & 0 \\ \sin\theta & \cos\theta & 0 & 0 \\ 0 & 0 & \cos\theta & -\sin\theta \\ 0 & 0 & \sin\theta & \cos\theta \end{pmatrix}$$

Note that there is only one parameter, θ.

$$\mathbf{4} = \mathbf{2} \oplus \mathbf{2} \quad SO(2)$$

An example, a representation of $SO(2) \otimes SO(2)$, which is a direct product of representations has the following matrix form:

$$R(\theta, \phi) = \begin{pmatrix} \cos\theta & -\sin\theta & 0 & 0 \\ \sin\theta & \cos\theta & 0 & 0 \\ 0 & 0 & \cos\phi & -\sin\phi \\ 0 & 0 & \sin\phi & \cos\phi \end{pmatrix}$$

Note that there are two parameters, θ and ϕ.

$$\mathbf{4} = \mathbf{2} \otimes \mathbf{2} \quad SO(2) \otimes SO(2)$$

function of the space coordinates, rather than constant quantities. In the simple example of Fig. 2.8, this means $\phi = \phi(\mathbf{x})$. To make the Lagrangian equation invariant, we need to introduce new terms to it, and these terms correspond to interactions. These theories are called *gauge* theories.

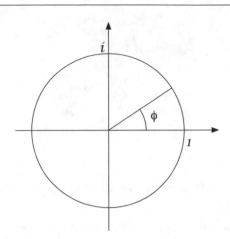

Fig. 2.8 Phase transformations depend on one $\mathbb{R}$eal parameter, ϕ. They act on complex functions $\psi(x) \rightarrow \psi'(x) = e^{i\phi}\psi(x)$ multiplying them by $e^{i\phi}$. They form a unitary and abelian group called U(1). If the parameter ϕ is a constant, we have a *global* transformation

2.8 Problems

2.1 A *heavy-flavoured B-meson* is a particle with a mass of approximately 5 GeV/c^2. For what values of momentum is its γ factor $\gamma \geq 3$.

2.2 Three spaceships fly in triangular formation at 100 km from each other at a constant speed of 0.8 c with respect to the space station Alpha, which can be considered to be an inertial system with good approximation. How is the radio communication among the three spaceships affected?

2.3 In the example in Chap. 2, a ^{90}Sr source produces β rays, which are electrons, with a kinetic energy distribution which extends up to 546 keV. This is called the *end-point* of the spectrum. What is the corresponding relativistic electron velocity?

2.4 A particle with mass of 125 GeV/c^2 decays into two γ's. Calculate the energy of the two gammas and their relative direction in the reference frame where the initial particle is at rest.

2.5 Prove that the space–time interval $ds^2 = c^2t^2 - dx^2 - dy^2 - dz^2$ is invariant by Lorentz transformations.

2.6 A photon can be described as a plane wave $A(\vec{x}, t) = A_0 \sin(\vec{k}\vec{x} - \omega t)$ with phase velocity $\frac{\omega}{|\vec{k}|} = c$. The angular frequency is related to the frequency by $\omega = 2\pi\nu$. The photon can also be treated as a massless particle, with 4-momentum $(E_\gamma/c, E_\gamma, 0, 0)$. The relation between energy and frequency is given to be $E_\gamma = h\nu$, where h is Planck's constant, which will be introduced in the next chapter. Show that the Doppler effect formulas (Eq. (2.55)) can be obtained by Lorentz-transforming the photon 4-momentum, and using $E_\gamma = h\nu$.

2.9   Solutions

Solution 2.1   We can use more than one formula as a starting point: the definition of momentum $p = \gamma\beta mc$ but we see that this depends on β as well. The modulus of the energy–momentum 4-vector is a good alternative: $E^2 - p^2c^2 = m^2c^4$; but, we also know that $E = \gamma mc^2$, and the formula only depends on p and m:

$$E^2 = m^2c^4 + p^2c^2 = \gamma^2 m^2 c^4$$

$$\gamma^2 = \frac{m^2c^4 + p^2c^2}{m^2c^4} = 1 + \frac{p^2}{m^2c^2}$$

$$p^2 = (\gamma^2 - 1)m^2c^2$$

Setting $\gamma = 3$, we have $p \geq 2\sqrt{2}mc = 2 \times 1.414 \times 5\,\mathrm{GeV/c^2} \times c = 14.1\,\mathrm{GeV/c}$.

Solution to 2.2   The spaceships are flying in formation so we can find a reference frame where the three spaceships are at a constant separation distance from each other. This frame is moving at a constant speed with respect to an inertial reference frame, and therefore it is in turn an inertial reference frame. The radio communications among these ships occur at the speed of light in vacuum, which is the same in all inertial reference frames. As the relative distance does not change, no Doppler effect is present.

Solution to 2.3   From Eq. (2.51):

$$E_k = (\gamma - 1)mc^2 \Rightarrow \gamma = 1 + \frac{E_k}{m_e c^2} = 1 + \frac{546}{511} = 2.068$$

From Eq. (2.26):

$$\gamma = \sqrt{\frac{1}{1 - \beta^2}} \Rightarrow \beta = \sqrt{1 - 1/\gamma^2} = \sqrt{1 - \frac{1}{4.28}} = 0.48$$

With a proper relativistic calculation, the electrons from ^{90}Sr have a maximum speed which is about half of the speed of light.

Solution to 2.4   In the reference frame where the initial particle is at rest, its total energy is $E = mc^2$. The initial momentum is zero, so also in the final state the total momentum must be zero. Therefore, the two photons must be emitted along the same straight line, but in opposite directions and with the same modulus of

3-momentum. For massless particles, $E = |p| c$ so the two photons have also the same energy:

$$p_{\gamma 1} = p_{\gamma 2}; \quad E_{\gamma 1} = E_{\gamma 2} = p_\gamma c$$

The invariant mass of the initial state is $E = 125\,\text{GeV}$. In the final state:

$$E' = |(E_\gamma, p_\gamma, 0, 0) + (E_\gamma, -p_\gamma, 0, 0)| = 2\,E_\gamma$$

$E' = E$ and therefore $E_\gamma = 1/2 \cdot 125 = 62.5\,\text{GeV}$.

Solution to 2.5   It is important to state the problem correctly. In this case, we cannot use trivially the time dilation and length contraction formulae, because they have implicitly built-in the hypothesis that the time interval is measured at the same place and the length is measured at the same time. Here, we have two independent events: $A = (ct_a, x_a, y_a, z_a)$ and $B = (ct_b, x_b, y_b, z_b)$. We can safely assume $y_a = y_b$ and $z_a = z_b$.

$$\Delta s^2 = c^2 (t_b - t_a)^2 - (x_b - x_a)^2$$

$$(\Delta s^2)' = c^2 (t_b' - t_a')^2 - (x_b' - x_a')^2$$

and we use the Lorentz transformations equation to express the primed quantities as a function of the coordinates in the non-primed reference frame. We can set $c(t_b - t_a) = a$ and $(x_b - x_a) = b$. The expression we obtain is

$$(\Delta s')^2 = \gamma^2 [a - (u/c)b]^2 - \gamma^2 [b - (u/c)a]^2$$

Developing the squares, we obtain

$$(\Delta s')^2 = \gamma^2 \left[a^2 - (u/c)^2 a^2 - b^2 + (u/c)^2 b^2 \right]$$

$$= \gamma^2 \left[a^2 (1 - (u/c)^2) - b^2 (1 - (u/c)^2) \right]$$

Recalling that $1 - \frac{u^2}{c^2} = \frac{1}{\gamma^2}$, we obtain:

$$(\Delta s')^2 = a^2 - b^2 = c^2 (t_b - t_a)^2 - (x_b - x_a)^2 = (\Delta s)^2$$

and therefore the *interval* is a Lorentz-invariant quantity.

Solution to 2.6   We assume that the reference frame is moving with velocity $\vec{u} = (u, 0, 0)$ in the same direction as the photon, which we assume along the $\hat{x}$ axis.

The photon 4-momentum is

$$\mathbf{p}_\gamma = (E_\gamma/c, E_\gamma, 0, 0) = (pc, p, 0, 0);$$

$$p'_x = \gamma \left(p_x - \frac{u}{c^2} E \right) \; ; \quad E = c\,p$$

$$p'_x = \gamma \left(p_x - \frac{u}{c} p_x \right) = p_x \frac{1 - u/c}{\sqrt{1 - (u/c)^2}}$$

Renaming $\beta = u/c$, we have

$$p'_x = p_x \sqrt{\frac{(1 - \beta)^2}{1 - \beta^2}} = p_x \sqrt{\frac{(1 - \beta)(1 - \beta)}{(1 + \beta)(1 - \beta)}} = p_x \sqrt{\frac{(1 - \beta)}{(1 + \beta)}}$$

The photon described in the initial frame has a frequency ν such that $p = h\nu/c$; in the primed reference frame, the photon has a frequency $\nu' = cp'/h$. So, replacing momentum with frequency we have

$$\nu' = \nu \sqrt{\frac{(1 - \beta)}{(1 + \beta)}} \; , \quad \text{which is the Doppler effect formula, Eq. (2.55).}$$

Bibliography and Further Reading

W.G. Dixon, *Special Relativity* (Cambridge University Press, Cambridge, 1978)
V. Faraoni, *Special Relativity* (Springer, Berlin, 2013)
H. Goldstein, *Classical Mechanics* (Addison-Wesley, Reading, 1980)
S.P. Puri, *Special Theory of Relativity* (Pearson, New Delhi/Dorling Kindersley, London, 2013)
W. Rindler, *Introduction to Special Relativity*, 2nd edn. (Oxford University Press, Oxford, 1991)
W.G.B. Rosser, *Introductory Special Relativity* (Taylor & Francis, New York, 1991)
J. Rosen, *Symmetry Rules* (Springer, Berlin, 2008)

Chapter 3
Essential Quantum Mechanics

3.1 Introduction

Before we continue, we must explain one other feature of radioactive decays: they produce α and γ particles of a well-defined energy, which are known as *spectral lines* (Fig. 3.1). This is very similar to what is observed in atomic physics: the light emitted by atoms, e.g. by a neon tube, has a sharply defined set of colour lines, which correspond to precise wavelengths of the light (Figs. 3.2).

The wavelength is a measure of the energy of the light, and the lines in the light spectrum emitted by a sample are the "fingerprints" of the elements contained in the sample. The analysis of light emission is effectively used to detect the presence of that element. The same holds true for γ-emitting nuclides: the set of spectral lines identifies uniquely the nuclide. In classical physics, the energy emitted by an atom is not constrained to a single value, or to a discrete set of values. An orbiting electron would quickly lose all its energy and fall into the nucleus. So, a different mathematical and physical framework is needed to describe the behaviour of small objects, like atoms and nuclei. In our daily routine, we are used to dealing with extended objects, which are mostly defined by their average values. We can define, e.g. the temperature, or the centre of mass of an extended body, or its inertial tensor. From these quantities, which are based on averages, we have extrapolated the properties of idealised point particles. This extrapolation was not always correct. When comparing with the real "small size particles", it was realised that they are not described by the same formulae and a new theory, Quantum Mechanics, has to be used instead of classical physics. There are many books on Quantum Mechanics which cover the topic extensively; a very small selection is reported in this chapter's

© Springer Nature Switzerland AG 2018
S. D'Auria, *Introduction to Nuclear and Particle Physics*,
Undergraduate Lecture Notes in Physics,
https://doi.org/10.1007/978-3-319-93855-4_3

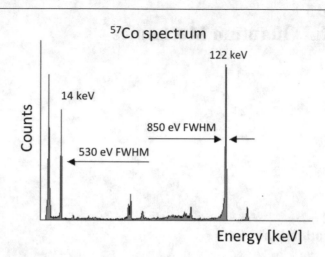

Fig. 3.1 Distribution of the gamma ray energy emitted by a ^{57}Co source and detected with a CdTe solid-state detector. Gamma rays are emitted in nuclear transitions (Courtesy Amptek, Inc., www.Amptek.com. The Full width at half maximum (FWHM) measures the detector resolution)

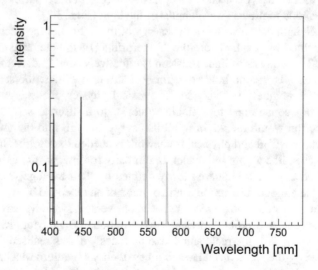

Fig. 3.2 A mercury vapour discharge lamp emits light with a *spectrum* which is not a continuum, as for incandescent lamps, but has a characteristic distribution of discrete lines. These lines correspond to the difference between energy levels in the atoms

bibliography. In order to make this book more self-consistent, the main ideas of quantum mechanics are included in this journey. Once again, we'll avoid a historical approach, which would start from the black body radiation. We can simply state the rules of quantum mechanics as coming from first principle, or a different kind of mathematical framework.

3.2   Principles of Quantum Physics

In quantum mechanics, we assume that all possible states of a given a physical system correspond to "vectors" of modulus one in some vector space. This space can be finite dimensional, or infinite dimensional and is called a *Hilbert space*.[1] An example of the first is the set of all numbers in the plane xy which lie on the edge of the unit circle $x^2 + y^2 = 1$; a base of this linear space is the set of the two vectors $(0, 1)$ and $(1, 0)$. An example of an infinite-dimensional vector space is the set of all functions of the coordinates which have the property that the square of the function, when integrated over its domain, results to a finite number. This function, which is called a wave function $\psi(\vec{x})$, can be expanded in a (infinite) Fourier series, i.e. it can be expanded in a base of this vector space. One representation does not exclude the other: we can have cases when a system is described by a two-component wave function: $\psi(\vec{x})(a(0, 1)+b(1, 0))$, with $a^2+b^2 = 1$ to preserve the unimodular property. This two-component function describes a *spin* $1/2$ particle, as will be clearer later. In quantum mechanics, there is, in general, less freedom than in classical mechanics: the possible states are often discrete and the really *observable* quantity, which can be measured and calculated, is the probability that a transition between two states takes place.

The evolution of a system is always represented by a linear operator acting in the appropriate Hilbert space. The transition probability between two states is given by the square of an amplitude value, and the result of a measurement is given by the mean value of the corresponding linear operator. The existence of discrete set of possible states, e.g. the energy levels of an atom, explains the discrete set of electromagnetic quanta, or photons, that are emitted. The spectral lines correspond to energy difference between allowed energy levels, as shown in Fig. 3.3.

The most important quantity in quantum mechanics is Planck's constant[2]:

$$h = 4.135667662(25) \times 10^{-15} \, \text{eV} \times \text{s, or } 4.13 \, \text{eV} \times \text{fs} \tag{3.1}$$

The number in parenthesis indicates the experimental error on the neighbouring two digits. This quantity has dimensions of an *action* = energy $\times$ time and is extremely small, even in our special units of electron Volt. This explains why we are not

---

[1] After David Hilbert (1862–1943), mathematician, was born in Konigsberg, Prussia (now Russia). He was professor in Göttingen, Germany.

[2] After Max Planck, 1858–1947, from Germany, who was awarded the Nobel prize in 1918 for his discovery of energy quanta.

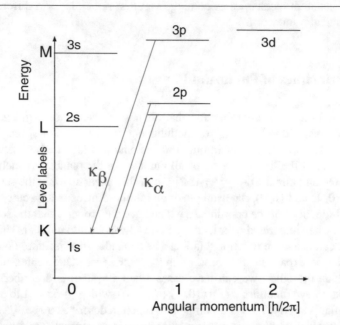

Fig. 3.3 In an atom, electrons can occupy only discrete energy levels. Electromagnetic radiation, which can be light or X-rays, is emitted when electrons make a transition from a level of higher-energy to a lower-energy level. The radiation is emitted with a single-energy value, corresponding to the energy difference between the two levels. Some rules select the permitted transition (*selection rules*). The transition probability and the energy difference can be exactly calculated with quantum mechanics

familiar with the quantum mechanical world: in most of the physics in everyday life, we can assume that this constant is zero. In this case, we obtain the laws of classical physics. The experimental value of h sets the scale, in terms of energy and time characteristics of a system, when Quantum Mechanics has to be used. In other words, for macroscopic quantities entering everyday experience we make an approximation, which no longer holds true for some systems, like atoms nuclei and elementary particles, which are *intrinsically small*. The quantum scale is set by the experimental value of the Planck's constant h. A first example of use of Planck's constant is to link the particle and wave description of intrinsically small objects: the energy of monochromatic light, of a single frequency ν, is an integer multiple of $h\nu$, or in other terms, a beam of monochromatic light is a flux of photons γ, each of energy

$$\lambda\nu = c; \quad E = h\nu = \frac{hc}{\lambda}. \tag{3.2}$$

Conversely, a beam of electrons of momentum p_e is described by a wave function ψ_e with wavelength λ_e and frequency ν_e which are given by the De Broglie formula:

$$\lambda_e = h/p_e \; ; \quad \nu_e = E_e/h \; . \tag{3.3}$$

A direct consequence of the wave description of a quantum particle is the un- certainty relations: we cannot measure both momentum and position with infinite precision, simply because they are related to each other by a Fourier trans- form.

When we observe the motion of a ball, or any extended object, our observation typically involves light to illuminate the object. Light does not interfere with the motion in any visible way. For small or fundamental objects, like electrons and protons, light or other fundamental objects is our only available probe; any light which is used to observe them, by definition of observation, interacts with the electron or the proton and changes their motion, because the energy of the photons is comparable with the energy of the object that they "observe" or interact with. We can no longer separate in a clear way the object under study from the observer.

Quantum Mechanics can be used with or without special relativity, and thus we have Q.M. or Relativistic Quantum Mechanics (RQM). In addition, a *second quantisation* formalism, also known as *canonical quantisation* in quantum field theory, describes systems of many quantum particles.

For the purposes of this book, we need to use just a few concepts from Q.M.:

- The state of a system is represented by a unitary vector in an abstract space. Sometimes, this is represented by a so-called *ket vector* $|p, s, l, m\rangle$ according to Dirac's notation $|\rangle$. Linear operators transform a vector of this space into another vector of the same space. An eigenvector of a linear operator is a vector that is left unchanged by the operator, up to a multiplication factor. As an example, rotations around an axis leave unchanged all the vectors which are parallel to the rotation axis. Some of these operators correspond to physically measurable quantities. The theory can predict the set of available states, or vectors, as eigenstates of physical operators in this space. The corresponding eigenvalues are the *quantum numbers*, and they are used to label a state, e.g. p, l, m, s in the *ket* notation above.

Paul A.M. Dirac (1902–1984) Nobel prize 1933, "for the discovery of new productive forms of atomic theory". He wrote the equation that reconciled quantum theory with special relativity, predicting the existence of anti- matter and explaining the electron spin.

Louis De Broglie, France 1892–1987, was awarded the Nobel prize in 1929 for the wave theory of the matter, which he formulated in his Ph.D. thesis.

- The transition from a state to another can be predicted only in terms of probability of such transition to occur.
- If a transition between the same initial and final states can occur with different intermediate processes, the total probability is given by the square of the sum of the amplitudes of individual processes: all superpositions are linear.
- There are physical quantities, which are mathematically represented by *conjugate variables* and cannot be measured precisely at the same time, not only experimentally, but also from a theoretical point of view. This is Heisenberg's *uncertainty principle*[3]:

$$\Delta x \, \Delta p_x \geq h/(4\pi) = \hbar/2; \tag{3.4}$$

$$\Delta y \, \Delta p_y \geq h/(4\pi) = \hbar/2; \tag{3.5}$$

$$\Delta z \, \Delta p_z \geq h/(4\pi) = \hbar/2; \tag{3.6}$$

$$\Delta E \, \Delta t \geq h/(4\pi) = \hbar/2 \tag{3.7}$$

where $\vec{x}$ and $\vec{p}$ indicate position and momentum, while E and t indicate energy and time, respectively.
- Planck's (3.2) and De Broglie's (3.3) relations, which link energy and momentum to frequency and wavelength, for photons and massive particles, respectively:

$$\lambda_\gamma \nu_\gamma = c; \quad E_\gamma = h\nu_\gamma = \frac{hc}{\lambda_\gamma}; \qquad\qquad p_\gamma = \frac{h\nu_\gamma}{c}$$

$$\nu_e = \frac{E_e}{h} ; \qquad\qquad \lambda_e = \frac{h}{p_e}$$

The *reduced Planck's constant* $h/(2\pi)$ is indicated with $\hbar = 0.658$ eV fs (electron Volt femtosecond).

3.3 Spin

Spin is peculiar to quantum theory and has no analogy in classical physics. It was introduced by Uhlenbeck and Goudsmith in 1926 to explain the emission lines of the hydrogen atom. It helps clarifying if we first mention the quantum behaviour of angular momentum, then introduce spin as an *intrinsic* angular momentum of particles. Experiments where atoms are deflected with a magnetic field have shown

---

[3]After Werner Heisenberg, 1901–1976, who was awarded the Nobel prize in 1932 for the creation of quantum mechanics.

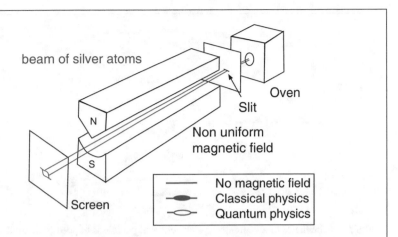

Fig. 3.4 The Stern and Gerlach apparatus demonstrates the quantum behaviour of the magnetic moment of atoms. In a non-uniform magnetic field, the beam of silver atoms is split into two lines, it is not just broadened, as classical physics predicted. The experiment was performed in 1922 in Frankfurt, Germany, by Walter Gerlach (1889–1979) and Otto Stern (1888–1969), who later was awarded the Nobel prize for physics in 1943. Gerlach was not mentioned, because of his leading role in nuclear research in Nazi Germany. Gerlach performed the conclusive experiment, and sent the result as a postcard to Niels Bohr. It should be noted that the S-G experiment did not discover the spin, but the quantum behaviour of magnetic moments of atomic particles. The original Stern–Gerlach experiment actually measured the *intrinsic* magnetic moment of the outer electrons of the Ag atoms, but it was not until later that this correct interpretation was found

that these particles have a magnetic moment, which behaves very differently from what expected from classical physics. The discrete states or energy values, which are typical of quantum systems, appear also when describing systems with angular momentum. Let's choose a direction in space: typically, this is the direction of an external magnetic field. If a quantum object, like particle or an atom, has a magnetic moment, its component along the chosen direction can only have a finite number of discrete values.

This was confirmed by Stern and Gerlach, in a famous experiment using a beam of silver atoms in a non-uniform magnetic field (Fig. 3.4). Classical physics linearly connects the magnetic moment μ of a macroscopic object to its angular momentum L:

$$\mu = \gamma L; \quad \gamma = \frac{q}{2m} \quad \text{(classical)} \tag{3.8}$$

The factor γ is the *gyromagnetic ratio*, q the electric charge of the object and m its mass. A quantised magnetic moment implies that also the angular momentum L can only have discrete values. Planck's constant has exactly the correct physical dimensions for an angular momentum, and therefore a natural unit for measuring

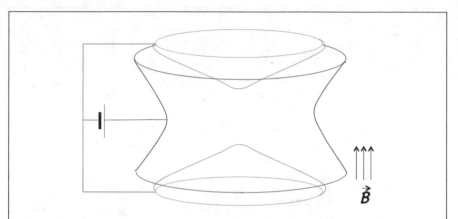

Fig. 3.5 Schematics of a Penning trap. The charged particle is trapped in the middle of this cavity, where a quadrupole electric field and a magnetic field are present. The Penning trap was first built by Hans G. Dehmelt (1922–2017) (Germany and the USA) who was awarded the Nobel prize in 1989 for the precise measurement of the electron magnetic moment. It is named after the Dutch physicist Frans Penning (1894–1953)

quantum angular momenta is $\hbar = h/2\pi$. In atoms, it comes as a natural extension of our macroscopic world to imagine the angular momentum of an electron as due to its orbit around the nucleus. When dealing with fundamental, point-like charged particles traveling in a straight line, we would expect no magnetic moment, and no angular momentum. In a Stern-Gerlach experiment using an electron beam, the Lorentz force would be much larger than the force on the magnetic moment due to the inhomogeneous magnetic field. However, in a *Penning trap* (Fig. 3.5).

It is possible to confine single electrons by means of electric and magnetic fields, and it is possible to measure very precisely that they have a magnetic moment. Single electrons carry an angular momentum, just like rotating objects. A required condition to rotate is to have a non-zero extension, while electrons, in all experiments so far, have demonstrated their point-like nature, with no measurable extension. These two pictures, a rotating, but point-like object, are difficult to reconcile. In order to overcome the difficulty, we need to discard the picture of a rotating object and recall that angular momentum is the quantity which is conserved when the mathematical description of the system, the Lagrangian equation, is invariant under rotations in space. A perfectly spherical object is invariant under rotations around its centre, and so is a mathematical point. They carry no "intrinsic" angular momentum, and they can be described by the position of the centre of mass. A rotating spherical extended object, just like a mathematical vector, is not invariant under rotations of the reference frame and has an intrinsic angular momentum. This is the key point to understand spin: the wave function description of most particles is not invariant under rotations of the reference frame. We should be careful now to specify that we have introduced two types of invariance: one is the invariance of the description of the system, which is called the Lagrangian function. Normally, for all systems this is the case: the Lagrangian is invariant under rotations, and the angular momentum is conserved. The second invariance is one of the wave functions: in order to conserve the angular momentum, the wave function has to transform under rotations, just like two vectors change their coordinates under rotation, but their scalar product remains invariant.

Let's now look at how to transform the wave functions, which will bring us to the explanation of the various discrete values of the spin. The experiments, e.g. electrons in a Penning trap, not only show that free electrons carry angular momentum but also that given any direction, the projection of the intrinsic angular momentum along this direction has only two possible values $\pm \hbar/2$. This behaviour can be described by a two-component wave function. We can be more familiar with a three-component wave function, e.g. the one which is needed to describe the electromagnetic field in classical physics: the magnetic field $\vec{B}$ has three components, and each of them is defined by a function of the coordinates:

$$\vec{B} = \vec{B}(\vec{x}) = (B_x(x, y, z), \ B_y(x, y, z), \ B_z(x, y, z)) \tag{3.9}$$

Relativistic notation requires a four-component vector, the electromagnetic vector potential, which is normally indicated with $\mathbf{A}(\mathbf{x})$ and describes photons. We know how rotations of the coordinate frame rotate these three-component or four-component fields, as rotations are a subgroup of the Lorentz group $\mathcal{L}$. However, we don't know, so far, how to rotate two-component fields. To do so, we need to take a step in group theory and recall that rotations form a group, SO(3). This group has representation in odd number of dimensions, so it is not trivial to use it in 2-D. However, there is a group which is "very similar" to it and has a representation in both even and odd dimensions.

This group is the set of complex unitary matrices 2×2 with unit determinant: SU(2). For any rotation matrix in our familiar three-dimensional space, we can find two corresponding 2×2 rotation matrices in SU(2), but for each rotation in SU(2) we can find only one rotation in SO(3); we use this matrix to "rotate" the two-component wave function. This category of wave functions correctly describes the quantum behaviour of particles which have intrinsic angular momentum $S = 1/2\hbar$. Wave functions with three components are called *vector wave functions*. They describe particles which have three possible values of angular momentum along any given direction: $S_z = 1\hbar$ or $S_z = 0$ or $S_z = -1\hbar$. Two-component wave functions are called *spinors* (Table 3.1). Quantum objects with no angular momentum are

Table 3.1 Relation between the value of the *spin* and the number of components of the wave function in non-relativistic quantum mechanics

Spin ($\hbar$)	Name	Components
0	Scalar	1
1/2	Spinor	2
1	Vector	3

described by a one-component wave function, which is a *scalar*. In relativistic quantum mechanics, the picture is more complicated, and it is outside the scope of this book. It can be understood in terms of representations of the Lorentz group: relativistic spinors have four components, but rotate differently from vectors, which also have four components. The two additional spinorial components describe the corresponding anti-particle. The existence of the positron was predicted by the Dirac's equation requiring a four-component rather than two-component spinor. Electrons, protons and neutrons have spin 1/2, photons have spin 1 and alpha particles have spin zero.

As it is more a "geometrical property", spin is conserved in all interactions and is linearly added when two or more particles form a bound state: an electron and a positron can combine and form two different bound states, called *positronium*: a spin zero (*para-*) or a spin one (*ortho-positronium*), with different behaviours; a proton and a neutron only form a spin-1 bound state, the *deuteron*, the α particles are bound states of two protons and two neutrons and have spin zero.

3.4 Spin and Magnetic Moment

Correlation between angular moment L and magnetic moment is straightforward in classical physics, by definition of the magnetic moment. For the spin S, Eq. (3.8) has to be modified with a numerical factor, which is normally indicated with g.

$$\mu = \gamma S = g \frac{q}{2m} S . \tag{3.10}$$

This number g is predicted to be $g = 2$ for spin $1/2\hbar$ particles, with some corrections, which can be calculated with quantum electrodynamics. All the ordinary stable or long-lived particles, electrons, protons, neutrons and *muons*, particles which we shall meet later, have *spin* $= 1/2\hbar$. The g-value is one of the best measured quantities:

$$g_e = 2.0023193043618 \ (5) \tag{3.11}$$

$$g_\mu = 2.0023318418 \ (13) \tag{3.12}$$

$$g_p = 5.585694702 \ (17) \tag{3.13}$$

where the number in parenthesis are the experimental errors on the last digits. Penning traps and particle accelerators are used for these measurements. It is important to notice that protons have a g-factor which is considerably different from the predicted value of two. The reason for this is their composite nature: protons and neutrons are made of more fundamental particles, partons or *quarks*, so they are a system of particles. For the same reason, also neutrons, which are electrically neutral, have a non-zero magnetic moment:

$$g_n = 3.82608545 \ (90) \tag{3.14}$$

Any deviation of the g-factor from the calculated value, for electrons or muons, would be an indication of compositeness and substructure.

Magnetic moments are measured in terms of *magnetons*:

$$\mu_e = g_e \frac{e}{2m_e} S = g_e \frac{e\hbar}{2m_e\hbar} S = g_e \frac{\mu_B}{\hbar} S \qquad (3.15)$$

where μ_B is called *Bohr's magneton*, while for protons and nuclei the *nuclear magneton* μ_N is used, where the electron mass is replaced with the proton mass:

$$\mu_B = \frac{e\hbar}{2m_e}; \quad \mu_N = \frac{e\hbar}{2m_p} \qquad (3.16)$$

Alpha particles have spin zero and also have zero magnetic moment.

Numerical values of the Bohr and Nuclear magnetons:

$$\mu_B = 5.7883818012(26) \times 10^{11} \text{ MeV T}^{-1}$$

$$\mu_N = 3.1524512550(15) \times 10^{14} \text{ MeV T}^{-1}$$

3.5 Problems

3.1 A discharge lamp operated with sodium vapour emits a characteristic yellow light, and is typically used in street lights. The wavelength of the two yellow lines is $\lambda_1 = 589.0$ nm and $\lambda_2 = 589.6$ nm. Calculate the energy in eV of these photons emitted by sodium, taking into account that the required precision is 10^{-3}.

3.2 The cathode ray tube, which was used in TV sets (Fig. 3.6), accelerates the electrons emitted by a filament with a voltage of about 2 kV. Are these electrons relativistic? Calculate the wavelength of these electrons when they reach the screen and compare it with the wavelength of visible light. The electron mass is $m_e = 511 \text{ keV/c}^2$.

3.3 We have a *photoelectric effect* when electrons are extracted from a metal surface to the vacuum, when the surface is illuminated by light.

We know from other measurements that to extract an electron from a metal surface a minimum energy E_w is required. This energy is called *work function*. Using Planck's formulas linking the energy of a photon to its wavelength,

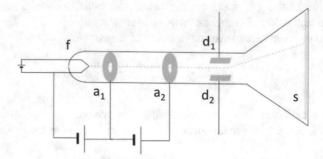

Fig. 3.6 Schematics of a cathode ray tube. The glass tube keeps all components in vacuum. Electrons are emitted by a filament (f) and accelerated by the electrodes e_1 and e_2. Other electrodes (d_1, d_2) or magnetic fields are used to bend the electron beam and to focus it on the screen (s), which is covered with an electro luminescent material

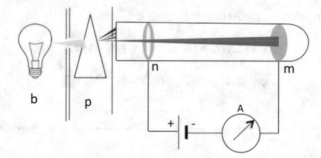

Fig. 3.7 Schematics of an apparatus to measure the photoelectric effect in metals. Light is selected with a prism (p) and directed towards a glass vacuum tube with metal electrodes inside. The electrode (m) emits electrons if conditions on the wavelength (λ) are satisfied. An accelerating potential between (m) and the electrode (n) attracts the photo electrons towards (n), and a current is measured by the nano-amperometre (A)

The explanation of the photoelectric effect was due to Albert Einstein (1879–1955). He was awarded the Nobel prize in 1921 for this, rather than for the relativity.

---

describe what is expected when a copper or a lithium electrode in vacuum is illuminated with monochromatic light, as a function of the wavelength, and at constant light intensity. We assume that there is another electrode at relatively positive voltage in the same vacuum tube, as in Fig. 3.7. For copper $E_w = 4.53$ eV, for lithium $E_w = 2.9$ eV.

3.6 Solutions

Solution to 3.1   Using Planck's equations:

$$E_{\gamma 1} = \frac{hc}{\lambda} = \frac{4.135667 \times 10^{-15} \text{ eV s} \cdot 2.99792 \times 10^8 \text{ m/s}}{589.0 \times 10^{-9} \text{ m}} = 2.1050 \text{ eV}$$

$$E_{\gamma 2} = \frac{hc}{\lambda} = \frac{4.135667 \times 10^{-15} \text{ eV s} \cdot 2.99792 \times 10^8 \text{ m/s}}{589.6 \times 10^{-9} \text{ m}} = 2.1029 \text{ eV}$$

The difference between these two energy levels of the sodium atoms is only 2.1 meV. It is due to the *spin–orbit* coupling of the electron, which is the interaction between the magnetic moment due to the electron orbital angular momentum L and the magnetic moment due to the electron spin S.

Solution to 3.2

$$\gamma_v = 1 + \frac{E_k}{mc^2} = 1 + \frac{2 \text{ keV}}{511 \text{ keV}} = 1.00391 \approx 1$$

So, we can use in first approximation non-relativistic formulae.

$$p_e = \sqrt{2m_e E_k} = \sqrt{2 \cdot 511 \cdot 2 \text{ keV}^2/c^2} = 2\sqrt{511} = 45.2 \text{ keV}/c$$

Using De Broglie's equation (3.3),

$$\lambda = \frac{h}{p_e} = \frac{4.13 \times 10^{-18} \text{ keV} \cdot \text{s} \cdot 3 \times 10^8 \text{ m/s}}{45.2 \text{ keV}} = 27.4 \text{ pm}$$

This wavelength is four orders of magnitude smaller than the wavelength of the visible light. The advantage of electron microscopes is in this small wavelength compared to light. They are not limited by diffraction in most cases.

Solution to 3.3   Photons of monochromatic light have all the same energy, whose value is given by $E_\gamma = hc/\lambda$. If this energy is lower than the work function E_w, electrons cannot be emitted and, therefore, no current circulates in the device. In case of copper cathode, for energies larger than 4.53 eV, corresponding to a wavelength

$$\lambda = \frac{hc}{E_w} = \frac{4.136 \times 10^{-15} \text{ eV s} \cdot 2.998 \times 10^8 \text{ m/s}}{4.53 \text{ eV}} = 273 \text{ nm}$$

the current starts circulating in the device, with a constant intensity. For lithium, $\lambda = 427$ nm. This effect only occurs for light with a wavelength smaller than a threshold value, which depends on the metal. For copper this wavelength is in the ultraviolet range, for lithium it corresponds to the indigo colour. The effect is used in photomultipliers and other devices, to detect photons.

Bibliography and Further Reading

This chapter was added for consistency, because quantum mechanics is taught in other courses. Many readers and students may already be familiar with it and many high-quality books explain it in much more detail. A very partial and incomplete list is given below, where introductory books and innovative presentations are preceding more classical approaches to the topic.

B.H. Brandsen, C.J. Joachain *Quantum Mechanics* (Prentice Hall, Englewood Cliffs, 2000)
W. Greiner, *Quantum Mechanics – An Introduction* (Springer, Berlin, 2001)
P. Pereira, *Fundamentals of Quantum Physics* (Springer, Berlin, 2012)
L.E. Picasso, *Lectures in Quantum Mechanics* (Springer, Berlin, 2017)
J.J. Sakurai, J. Napolitano, *Modern Quantum Mechanics* (Pearson Education, New Delhi, 2013)
L. Susskind, A. Friedman, *Quantum Mechanics: The Theoretical Minimum* (Basic Books, New York, 2014)
K. Sundermeyer, *Symmetries in Fundamental Physics* (Springer, Berlin, 2014)

Chapter 4
Radioactive Decays

4.1 Introduction

Decays are a category of reaction where a particle, or a nucleus, transforms into two or more particles or nuclei. We have already introduced the α-, β- and γ-emitting decays. In the α decay (Fig. 4.1), the parent nucleus emits an α particle, which is a nucleus of $_2^4\text{He}$, and the resulting daughter nucleus has an atomic number which is two units lower and a mass number which is four units lower. In the β decay (Fig. 4.2), an electron or a positron is emitted, and the resulting daughter nucleus has an atomic number lower by one unit, while the mass number remains unchanged (Table 4.1).

The γ radiation is emitted by nuclei which are in an excited state and no change in their composition occurs. Typically, nuclei are in an excited state after α or β decays, so the γ radiation is associated with the other two.

In all decays, the 4-momentum is conserved. We recall that we indicate with A the total number of nucleons and with Z the number of protons, so the pair (A, Z) uniquely identifies a nuclide. For α decays, we have (Fig. 4.3)

$$(A, Z) \rightarrow (A - 4, Z - 2) + \alpha. \tag{4.1}$$

In the rest frame of the parent nucleus, the α particle and the daughter nucleus recoil back to back, with the same momentum. The energy states of nuclei have discrete values, and these two conditions explain why, for a given nuclide and a given daughter, the α particles emitted have all the same energy.

With the same conditions, the continuous spectrum of the $\beta^{\pm}$ decay seems to defy the conservation of 4-momentum:

$$\beta^{\pm}\text{decay } (A, Z) \rightarrow (A, Z \mp 1) + \beta^{\pm} \text{ (missing a term)} \tag{4.2}$$

© Springer Nature Switzerland AG 2018
S. D'Auria, *Introduction to Nuclear and Particle Physics*,
Undergraduate Lecture Notes in Physics,
https://doi.org/10.1007/978-3-319-93855-4_4

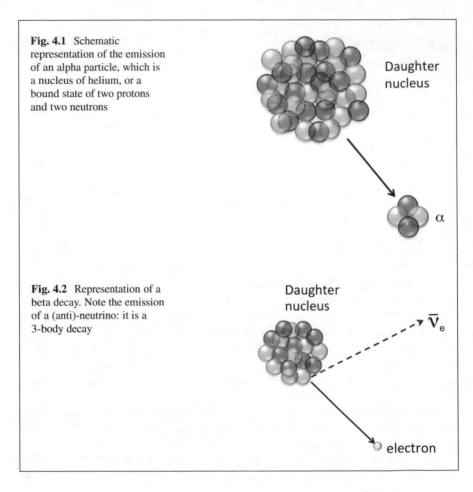

Fig. 4.1 Schematic representation of the emission of an alpha particle, which is a nucleus of helium, or a bound state of two protons and two neutrons

Daughter nucleus

α

Fig. 4.2 Representation of a beta decay. Note the emission of a (anti)-neutrino: it is a 3-body decay

Daughter nucleus

$\overline{\nu}_e$

electron

Table 4.1 Table of the nuclear transmutation due to α and β decays

Decay	Parent	Daughter
α	(A, Z)	$(A-4, Z-2)$
β^-	(A, Z)	$(A, Z+1)$
β^+	(A, Z)	$(A, Z-1)$
γ	(A, Z)	(A, Z)

   Also in this case, we would expect the electron or positron to recoil against the nucleus and have only one value of momentum, or just a restricted range of momenta, to account for a natural "width", of quantum origin. This puzzle led Wolfgang Pauli to make the hypothesis that another neutral particle is emitted and is not detected experimentally. He called the particle *neutron*, but the discovery of what we now call a neutron led Enrico Fermi to call this particle *neutrino*, for "the little neutral" particle. It is indicated with ν_e; the subscript e will be clearer later.

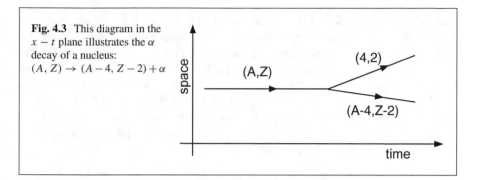

Fig. 4.3 This diagram in the $x - t$ plane illustrates the α decay of a nucleus: $(A, Z) \rightarrow (A - 4, Z - 2) + \alpha$

Only 30 years later, the neutrino was detected experimentally. The correct decay reaction is

$$\beta^{\pm}\text{decay } (A, Z) \rightarrow (A, Z \mp 1) + \beta^{\pm} + \nu_e \qquad (4.3)$$

In order for a decay to occur, the final state must be energetically allowed. The energy must be conserved, so the sum of total energy of the final state, including the kinetic energy, must be equal to the total energy of the initial particle or nucleus, but this will be the topic in the last chapter of this book.

4.2 The Laws of Radioactive Decay

Independently of the mechanism and type of decay, we'll introduce now some general considerations that apply to all types of decays and derive the law of the time evolution of radioactive decays.

The physics of decays is the same in every reference frame. There is one frame which makes calculations easy, and it is the frame where the initial particle is at rest.

We can have two pictures: one is the one-particle picture of the decay; we pick up one particle and after a time, which is random, we observe it to decay. We also have the opportunity to repeat the experiment N times, with N particle of identical type and nature; we see that the decay times are distributed according to a mathematical function.

In the single particle picture, the basic assumption we can make is that the probability of the decay to occur within the *small* time interval Δt is constant in time; it does not depend on time at all. Of course, the larger is the *small* time interval in which we observe, the larger is the probability dP to decay. It is natural to assume that, for small, but finite, intervals Δt, the probability to decay is just proportional to the time.

$$dP = \lambda dt = \frac{1}{\tau_0} dt \ . \qquad (4.4)$$

The constant λ must have physical dimensions $[T^{-1}]$, and it is called *decay rate*, while $1/\lambda = \tau_0$ is the *mean lifetime* or simply *lifetime*. Suppose we have N_0 radioactive nuclei at rest at time t_0. Then, Eq. (4.4) applies to each of them. We can recall the frequentist definition of probability, as the relative frequency of an event, i.e. the ratio between the number of times a certain event occurred and the number of times that it was possible that the event occurred, in the limit of a large number of tries. We observe for a *small* time Δt a very large number N_0 of initial nuclei; this is our number of tries, because each of them could decay. Out of those, we observe a small number N_d decay:

$$dP = \frac{N_d}{N_0} = \frac{\Delta t}{\tau_0} \tag{4.5}$$

for the number of decays N_d. We now want to know the variation of the number N of initial nuclei or particles: $\Delta N = -N_d$. Substituting in Eq. (4.5), we have

$$\frac{\Delta N}{\Delta t} = -\frac{N}{\tau_0} \tag{4.6}$$

(note the minus sign) which is valid at any time. At the limit of infinitesimal time interval $\Delta t \to dt$, we obtain

$$\frac{dN}{dt} = -\frac{N}{\tau_0}, \tag{4.7}$$

which we can consider as a differential equation for the number of nuclei, or particles, which are still undecayed at the time t, i.e. the function $N(t)$. The solution is

$$N(t) = N_0 e^{-t/\tau_0} \tag{4.8}$$

This formula fits extremely well the experimental measurements. An example is shown in Fig. 4.4

We can verify that

$$\frac{d}{dt}N(t) = -\frac{1}{\tau_0}N_0 e^{-t/\tau_0} = -\frac{N(t)}{\tau_0}$$

Suppose we have a radiation detector next to our radiation source (Fig. 4.5). In general, it will detect and count only a fraction ϵ of the radiation emitted due to an intrinsic detection efficiency and to geometrical acceptance. The counting rate R will be proportional to the *activity* $\mathcal{A}$ of the source:

$$R = \epsilon \mathcal{A} = \epsilon(-dN/dt) = \epsilon \lambda N(t)$$

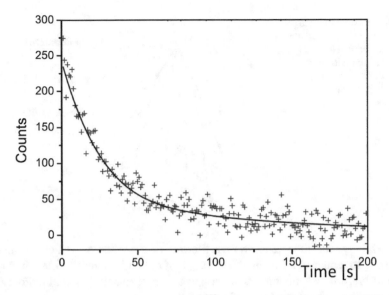

Fig. 4.4 Experimental decay curve of the nuclide ^{229}Rn, a beta emitter discovered in 2009. From D. Neidherr et al., Phys. Rev. Lett. 102 (2009)

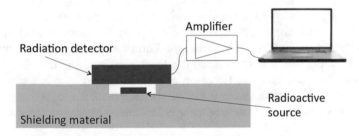

Fig. 4.5 A particle *detector* is used to measure the activity of a radioactive source. In this case, only about 50% of the emitted radiation can reach the detector. In addition, especially in case of γ radiation, some radiation may escape detection, contributing to inefficiency, which must be taken into account to calculate the real activity of the source

So, the activity of the source at time t is

$$\mathcal{A}(t) = \lambda N(t) = \frac{1}{\tau} N(t) = \frac{1}{\tau \epsilon} R(t) \tag{4.9}$$

The activity of a radioactive source (or material) is measured in *Becquerel (Bq)*, which corresponds to one decay per second. Its physical dimensions are $[T^{-1}]$. A radioactive source of 1 Bq, if measured with a perfect detector, will give a count rate of 1 Hz. Of course, the normal multiple and sub-multiples of this unit are also used. Another unit for the activity, which is not an SI unit, is the *Curie* (Ci), a practical unit which is defined from the activity of one gram of ^{226}Ra : 37 MBq= 1 mCi. From Eq. (4.8), it is evident that the activity of a radioactive source decreases

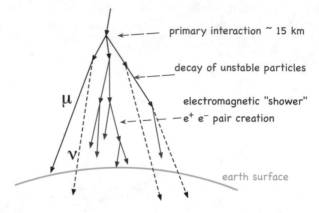

Fig. 4.6 Cosmic rays are high-energy particles coming from the space. They start interacting with the outer layers of the atmosphere, generating cosmic-ray showers

exponentially with time. The average lifetime of a nuclide or particle is the time occurring to reduce to a fraction $1/e = 0.368$ of the initial sample. Often, it is given the *half-life*, which is the time occurring to a sample to half its initial radioactivity. The two values are related by $\log(2)$.

So far, we have considered that the nuclei are always at rest with respect to the laboratory. This is not always the case for elementary particles, or exotic nuclei. An elementary particle which we call *muon* ($\mu^{\pm}$) has a mean lifetime of $2.197\,\mu s$ when it is at rest in the lab. It is produced copiously by cosmic rays at the top of the atmosphere, say at 30 to 15 km from the Earth surface (Fig. 4.6). At the speed of light, it would take $100\,\mu s$ to reach the Earth's surface, and this time is much larger than the muon lifetime. However, as time is dilated by relativistic effects, the muon lifetime, as measured in the laboratory reference frame, is increased by a factor $\gamma(v)$. So, we are constantly bombarded by muons from cosmic rays.

4.3   More on Radioactive Decays

The radioactive decay is a particular kind of nuclear reaction, where a species "A" decays into two or more *decay products* (B, C, D). Some typical examples are the alpha decay:

$$^{241}\text{Am} \rightarrow ^{237}\text{Np} + \alpha(5.49\,\text{MeV}) \tag{4.10}$$

a beta decay:

$$^{90}\text{Sr} \rightarrow ^{90}\text{Y} + \beta^- + \overline{\nu_e} \tag{4.11}$$

and a gamma decay:

$$^{99m}\text{Tc}^* \rightarrow ^{99}\text{Tc} + \gamma(140\,\text{keV}) \tag{4.12}$$

The gamma decay is rather a transition between two *isomers* of the same nuclide, with emission of energy. We have also mentioned an elementary particle, the muon $\mu^{\pm}$, it decays

$$\mu^- \rightarrow e^- + \nu_\mu + \overline{\nu_e} \tag{4.13}$$

with a mean lifetime of $2.197\,\mu s$.

There is one reference frame, where the initial particle, which is also called the "parent", is at rest. In this frame, the decay mean lifetime is minimum and is a characteristic of the decay.

For some nuclides, or elementary particles, two or more decay channels may be open: for example, a nucleus can undergo both α and β decay. We say that there is a *branching*, and we can write the probability for a nucleus to undergo α decay in the time dt as:

$$dP_\alpha = \lambda_\alpha dt \tag{4.14}$$

and we can write something similar for the β decay:

$$dP_\beta = \lambda_\beta dt \tag{4.15}$$

then the total probability to decay in either mode is

$$P_{tot} = (\lambda_\alpha + \lambda_\beta)dt \tag{4.16}$$

and therefore

$$\frac{dN}{dt} = -(\lambda_\alpha + \lambda_\beta)N \tag{4.17}$$

The *average lifetime* of the given nuclide has only one value:

$$\tau = \frac{1}{\lambda_\alpha + \lambda_\beta} \tag{4.18}$$

The α *Branching Ratio* ($\mathcal{BR}$) is the ratio of the number of alpha decays to the total number of decays in a given time:

$$\mathcal{BR}(\alpha) = \frac{\lambda_\alpha}{\lambda_\alpha + \lambda_\beta} \tag{4.19}$$

An example in nuclear physics is

$$^{213}\text{Bi} \rightarrow\, ^{213}\text{Po} + \beta^- + \overline{\nu}_e \qquad 1.423\,\text{MeV}\,(97.8\%) \tag{4.20}$$

$$^{213}\text{Bi} \rightarrow\, ^{209}\text{Tl} + \alpha \qquad 5.87\,\text{MeV}\,(2.20\%) \tag{4.21}$$

Table 4.2 The main decay modes of the particle Z^0, with the corresponding branching fractions

Decay	Mode	Fraction (%)
Γ_1	$e^+ e^-$	(3.363 ± 0.004)
Γ_2	$\mu^+ \mu^-$	(3.366 ± 0.007)
Γ_3	$\tau^+ \tau^-$	(3.367 ± 0.008)
Γ_4	$\ell^+ \ell^-$	(3.3658 ± 0.0023)
Γ_5	Invisible	(20.00 ± 0.06)
Γ_6	Hadrons	(69.91 ± 0.06)
Γ_9	$c\bar{c}$	(12.03 ± 0.21)
Γ_{10}	$b\bar{b}$	(15.12 ± 0.05)

As an example: the decay $Z^0 \rightarrow e^+ e^-$ occurs with a branching fraction of 3.36%. The branching fractions above add up to more than 100% because they are not exclusive: ℓ indicates any of the e, μ, τ particles, and $c\bar{c}$ and $b\bar{b}$ are also included in the decay to "hadrons". All these particles will be more familiar in Chap. 6. Data from the Particle Data Group

An example in particle physics is the decay of the particle Z^0, as shown in Table 4.2.

In a simple decay $A \rightarrow B + C$, the products B and C are stable. However, a decay product B can itself be a radioactive nuclide, or an unstable particle. In this case, we have a *chain* of decays: B can decay into products, which are themselves unstable, and so on. An example is the decay of radon, a naturally occurring gas which can be found in basements of some buildings.

$$^{222}\text{Rn} \rightarrow \alpha + ^{218}\text{Po} \rightarrow \alpha + ^{214}Pb \rightarrow \dots \tag{4.22}$$

which in turn is produced by decay of thorium, which is produced by Uranium decays:

$$^{230}\text{Th} \rightarrow \alpha + ^{226}\text{Ra} \rightarrow \alpha + ^{222}\text{Rn} \tag{4.23}$$

For α decays, there are only 4 possible chains of nuclides (Table 4.3): those whose mass number A is exactly divisible by 4 ($A = 4n$) and those whose number of nucleons is $A = 4n + 1$ or $A = 4n + 24$, or $A = 4n + 3$, where n is a positive integer. The corresponding series are called: thorium series ($A = 4n$), neptunium series, uranium series (Fig. 4.7) and actinium series ($A = 4n + 3$). The Neptunium series ($A = 4n + 1$) contains no nuclide with extremely large lifetime. The majority of the radioactive elements of the neptunium series are already extinct since their formation in a supernova, about five billion years ago.

Table 4.3 Table of the four possible series of α decay chains

Series	A	Final
Thorium	$A = 4n$	^{208}Pb
Neptunium	$A = 4n + 1$	^{205}Tl
Uranium	$A = 4n + 2$	^{206}Pb
Actinium	$A = 4n + 3$	^{207}Pb

The decay chains stop when they reach a stable nuclide, an isotope of lead or thallium

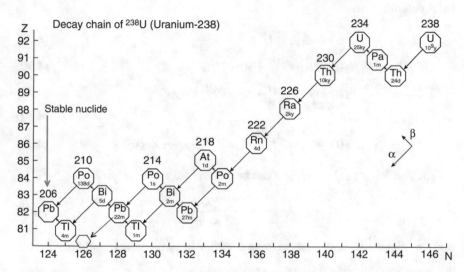

Fig. 4.7 The decay chain of ^{238}U is represented in the $N-Z$ plane. This representation of nuclides is called *Segrè chart*, from Emilio Segrè (1905–1989), who was awarded the Nobel prize in 1959 for the discovery of the anti-proton. N indicates the number of neutrons, $N + Z = A$. The α and β^- transitions are represented in an intuitive way. Figure from E. Segrè, "Nuclei and particles", W. Benjamin, Inc., 1965

Suppose to have the following reactions, where R_j indicates radioactive nuclides and $N_j(t)$ its corresponding number of nuclei present, or concentration at time t:

$$R_1 \rightarrow R_2 + A \tag{4.24}$$

$$R_2 \rightarrow R_3 + B \tag{4.25}$$

$$R_3 \rightarrow R_4 + C \tag{4.26}$$

The grand-parent radioactive nuclide R_1 follows a simple decay law $N_1 = N_1(0)e^{-\lambda_1 t}$, while the radioactive nuclide R_2 decays, but is also produced by the decay of R_1, so we can write

$$\frac{dN_2}{dt} = +\lambda_1 N_1 - \lambda_2 N_2 \tag{4.27}$$

and for nuclide R_3

$$\frac{dN_3}{dt} = +\lambda_2 N_2 - \lambda_3 N_3 \tag{4.28}$$

We have a chain of differential equations, which can be solved recursively. The general solution is reported below as a reference. It can be written in terms of sum of exponential decays, each term depends on the decay rate of all nuclides "above" in its genealogy tree.

$$N_1(t) = a_{11}e^{-\lambda_1 t} \tag{4.29}$$

$$N_2(t) = a_{21}e^{-\lambda_1 t} + a_{22}e^{-\lambda_2 t} \tag{4.30}$$

$$N_3(t) = a_{31}e^{-\lambda_1 t} + a_{32}e^{-\lambda_2 t} + a_{33}e^{-\lambda_3 t} \tag{4.31}$$

$$\cdots \tag{4.32}$$

$$N_k(t) = a_{k1}e^{-\lambda_1 t} + a_{k2}e^{-\lambda_2 t} + \cdots + a_{kk}e^{-\lambda_k t} \tag{4.33}$$

The coefficients a_{kj} with $k \neq j$ can be determined recursively:

$$a_{k,j} = a_{k-1,j}\frac{\lambda_{k-1}}{\lambda_k - \lambda_j} \tag{4.34}$$

and the "diagonal" coefficients must be determined by the initial conditions $N_k(0)$:

$$N_k(0) = a_{k1} + a_{k2} + \cdots + a_{kk} \tag{4.35}$$

A notable case is when only the radionuclide R_1 is initially present: $N_k(0) = 0$ for $k > 1$:

$$N_1(t) = N_1(0)e^{-\lambda_1 t} \tag{4.36}$$

$$N_2(t) = N_1(0)\frac{\lambda_1}{\lambda_2 - \lambda_1}(e^{-\lambda_1 t} - e^{-\lambda_2 t}) \tag{4.37}$$

$$N_3(t) = N_1(0)\lambda_1\lambda_2 \left(\frac{e^{-\lambda_1 t}}{(\lambda_2 - \lambda_1)(\lambda_2 - \lambda_1)} + \frac{e^{-\lambda_2 t}}{(\lambda_3 - \lambda_2)(\lambda_1 - \lambda_2)} \right.$$
$$\left. + \frac{e^{-\lambda_3 t}}{(\lambda_1 - \lambda_3)(\lambda_2 - \lambda_3)} \right) \tag{4.38}$$

If, at a certain point of the decay chain, one nuclide, say R_s, has a lifetime much larger than the others (Figs. 4.8 and 4.9), we can have a notable case of equilibrium, which is called *transient equilibrium*, and it occurs when all "younger generations" decay with substantially the same decay constant as the nuclide with long lifetime R_s above them in the chain. The term *"secular equilibrium"* indicates that, in addition, we observe the decay on a timescale where we can approximate

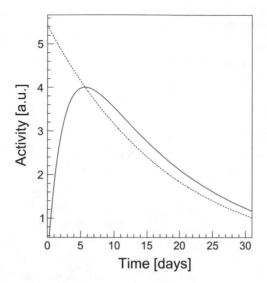

Fig. 4.8 An example of transient equilibrium, as it shows in the activity of a sample, which is initially composed of 100% Ba. The decay chain is: ^{140}Ba $\rightarrow^{140}$ La $\rightarrow^{140}$ Ce. The activity due to the parent is shown with a dashed line, and the activity of the daughter is represented by a solid line. The daughter nuclide ^{140}La has a lifetime 8 times smaller than the parent ^{140}Ba, and after a maximum the two curves follow each other. The activity of the daughter nuclide is larger than the activity of the parent

$e^{-\lambda_s t} = 1$. In this case for any R_t with $t > s$, we have

$$\frac{N_t}{N_s} = \frac{\lambda_s}{\lambda_t} \qquad (4.39)$$

Uranium ores, for instance, are in secular equilibrium.

Another important case is when a nuclide is continuously formed, not by decay but with a nuclear reaction. Examples of such cases are the formation ^{14}C in the atmosphere and the activation of material in a particle accelerator. The differential equation is

$$\frac{dN}{dt} = Q - \lambda N(t) . \qquad (4.40)$$

When the initial state is $N(0) = 0$, we have this solution:

$$N(t) = \frac{Q}{\lambda}(1 - e^{-\lambda t}) . \qquad (4.41)$$

The nuclide ^{14}C is continuously produced in the atmosphere by cosmic rays, with a reaction reported in Eq. (7.36). We can consider the concentration of ^{14}C in the

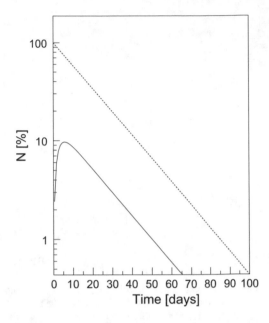

Fig. 4.9 The same decay as in Fig. 4.8 ^{140}Ba $\rightarrow ^{140}$La $\rightarrow ^{140}$Ce is used to illustrate the transient equilibrium, as in Eqs. (4.36) and (4.37). The fraction number of nuclides $N(t)/N_0$(Ba) existing at time t for Ba (dashed line) and La (solid line) is plotted as a function of time. After a transient, the two curves are almost parallel

atmospheric CO_2 to be constant in time: Eq. (4.41) for $t \rightarrow \infty$ gives $N(t) = Q/\lambda$. Corrections to the above equilibrium are due to human activities, like the 2055 nuclear explosions, which have increased the ^{14}C concentration; a peak of about twice the original value was reached in the mid of 1960s. Also, nuclear plants produce a negligible amount of ^{14}C, according to the reaction in Eq. (7.35). The use of fossil fuels, which are naturally depleted of ^{14}C by decay, dilutes its concentration in the atmosphere. The nuclear-test peak of [^{14}C] concentration in air has an exponential decay, with a mean lifetime of about 23 years, which is much shorter than the decay time. This is due to its gradual absorption, mainly by vegetation. The present concentration level is about the same as before the nuclear tests.

4.4 Age Determination with Radioactive Nuclides

When living organisms die, they stop exchanging CO_2 with the atmosphere. If we assume that ^{14}C concentration was constant in the past, knowing the decay time of ^{14}C and measuring the concentration of ^{14}C in the sample, we can measure the time since the sample has stopped exchanging carbon with the atmosphere (Fig. 4.10). We can solve the simple decay formula for the time interval Δt:

$$\Delta t = -\tau \ln \frac{N(t)}{N_0} = \tau \ln \frac{N_0}{N_{\text{sample}}} , \qquad (4.42)$$

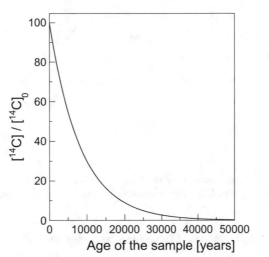

Fig. 4.10 Decay curve of ^{14}C, which is used for dating organic material. We measure the fraction of the radionuclide concentration with respect to the stable isotope $[^{14}C]/[^{12}C]$ in the sample. We then divide this value by the same fraction, measured in present living material, or by the reference value. This value is reported on the y-axis, as a function of time before present. This method was invented and published in 1946 by Willard F. Libby (1908–1980), from the USA, who taught at the University of Chicago and was awarded the Nobel prize in chemistry in 1960

where N_0 in this case is the concentration of ^{14}C in a sample exchanging CO_2 and $N(t)$ is the concentration of ^{14}C in our "historical" sample. ^{14}C decays β^- with a mean lifetime of 8267 years and an end-point energy of 156 keV. Precise dating with radiocarbon is sensitive to the natural variations of ^{14}C concentration in the atmosphere. Because of the peak due to nuclear tests, the standard concentration of ^{14}C is assumed as the value measured in 1950 and corresponds to a concentration of 1.30×10^{-12} with respect to all atoms of carbon. Radiocarbon dating is calibrated with increasing precision using information from tree rings and from analysing the air samples trapped in the ice layers in Greenland and Antarctica. Recent studies suggest that a considerable modulation in ^{14}C concentration occurred in a time interval of 30,000–45,000 years ago.

Other nuclides are used for longer time constants. In particular, the ratio between the concentrations of parent and daughter nuclides gives information on the time when a particular rock became solid. The principle is that since the solidification occurred, the daughter atoms stayed in place in the sample. The potassium–argon method uses ^{40}K, which has a half-life of 1.28×10^9 years; this is about 1.3 billion years, to be compared to the age of the Earth, which is 4.54 billion years. Therefore, this nuclide is useful for geological dating. It decays by *electron capture* to ^{40}Ar, which is chemically inert and escapes from the fluid magma, but remains trapped in solid rock.

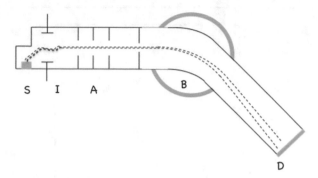

Fig. 4.11 Sketch of a mass spectrometer: (S) Sample, which is vaporised in vacuum, (I) ionisation, (A) accelerator, (B) magnetic field perpendicular to the drawing and (D) detector

4.5 Technical Aspects: The Mass Spectrometer

It would be intuitive to measure the concentration of a radionuclide for age deter-mination by measuring the activity of the sample. However, a simple calculation reveals that in most cases it is impractical or impossible to detect this radiation. A very powerful tool comes to our help: the *mass spectrometer* (Fig. 4.11). The sample is first pulverised, or purified with chemical methods. When heated in vacuum, its vaporised atoms are first ionised, then accelerated. The ion beam traverses an area with a magnetic field, and ions are deflected differently according to their different mass over charge ratio m/q. A detector at the end of the vacuum pipe is spatially segmented, to count the signals from single ions in different positions corresponding to different m/q values. This method allows us to measure the relative abundance of isotopes of the same chemical element present in the sample. It is also clear that this technique of dating with radioactive nuclides is destructive for the sample. It is clear that both the precision and the maximum age determination depend on the weight of the sample and on its chemical composition. Modern radiocarbon techniques can reach dating up to 70,000 years ago, while other nuclides are used for dating older samples.

4.6 Radioactive Iodine for Thyroid Cancer

This is a case study, to attract the reader's interest to a practical case. The numerical values used are representative of a real case, although in practice there are many other factors to be considered.

Thyroid cancer is cured with radioactive iodine-131, which accumulates in the thyroid much more than in other organs. The nuclide ^{131}I decays β^-, with an end-point electron energy of 606 keV. These β^- particles ionize the fluid, and the resulting ions damage the surrounding cancer cells. The daughter nuclide is formed in an excited state and emits a γ ray of energy $E_\gamma = 364\,\text{keV}$. The atomic weight of iodine is $A = 131$ and its half-life is 8.02 days.

1. What particles and nuclides are produced in this decay? (use the periodic table)
2. How many grams of iodine are needed to prepare a drink with initial activity 1.85 GBq? What would you use to precisely measure this quantity?
3. What would block this β radiation?
4. What type of radiation can be detected outside the patient's body?
5. Neglecting any disposal of iodine, how long the patient must be in isolation to see its thyroid radioactivity decreased to an acceptable level of 500 Bq?
6. 90% of the iodine is eliminated by the body in 3 days. Compare this time with the decay time above, and suggest the appropriate measures to the hospital.

Discussion

1. *What particles and nuclides are produced in the β^- decay of ^{131}I?*
   The β^- decay produces an electron with a continuum spectrum (as opposed to a single spectral line of monoenergetic particles). To conserve energy, another neutral particle, which escapes detection, must be emitted. This is the electron anti-neutrino $\overline{\nu}_e$. A neutron in the nucleus becomes a proton, thus the nuclide keeps the same mass number A and adds one to its atomic number, moving by one place to the right in the periodic table. To the right of iodine, we find Xenon, so the reaction is

$$^{131}I \to {}^{131}\text{Xe} + e^- + \overline{\nu}_e$$

2. *How many grams of ^{131}I are needed to prepare a drink with initial activity 1.8 GBq?*

$$t_{1/2}(^{131}\text{I}) = 8.82 \text{ days } = 8.02 \times 24 \times 3600 = 6.93 \times 10^5\text{s}.$$

$$\tau = t_{1/2}/\ln 2 \approx 1.44 \times t_{1/2} = 9.95 \times 10^5 \text{ s}$$

$$\text{activity } A = \lambda N(t) = \frac{N(t)}{\tau} = \frac{N_0 e^{-\frac{t}{\tau}}}{\tau}$$

We are interested to the number of nuclei for a given initial activity, i.e. at $t = 0$.

$$N_0 = \tau A = 9.98 \times 10^5 \times 1.8 \times 10^9 = 1.8 \times 10^{14}$$

We know that 6.022×10^{23} atoms (Avogadro's number) have a mass of A (mass number) grams. Therefore, in terms of mass: 1.8×10^{14} atoms of Iodine have a mass w:

$$w = \frac{1.8 \times 10^{14}}{6.022 \times 10^{23}} \times 131 \text{ g} = 39.1 \times 10^{-9} \text{ g} = 39.1 \text{ ng}$$

Modern analytical balances typically need samples of 1 mg or more, and have a sensitivity of about 100 ng. To measure this small quantity of 40 ng, we cannot use a balance; the mass is measured indirectly by measuring the activity, with using a radiation detector and a rate metre.

3. *What would block this β radiation?*

The β radiation of iodine-131 has an end-point of 606 keV and is blocked by a few millimetres of living tissue, which is mostly water. Incidentally, this is the very reason why this kind of radiation destroys cancer cells: beta radiation delivers energy to the cancer cells or to their surrounding fluid and the resulting ions damage the cells. The thyroid is under a few millimetres of neck skin.

   As a result, the beta radiation is blocked primarily by the cancer cells and by the surrounding tissue. The range of 606 keV beta particles in air is of the order of 30–50 cm.

4. *What type of radiation can be detected outside the patient's body?*

To a first approximation, only the γ radiation emitted by the daughter nucleus can be detected from outside the patient body. What follows is a more detailed explanation. The xenon nucleus, which is formed by the beta decay of the Iodine, finds itself surrounded by the electron cloud of the iodine, which has one electron less. An electron fills the lowest energy level of Xe, which is lower than the corresponding lowest level of Iodine. The atom emits X-rays with energies up to the κ_β energy of Xe, which is 33.6 keV. The γ radiation from the electromagnetic decay of the excited Xe nucleus has an energy of 364 keV. The passage of X-rays through the matter will be introduced in the next chapter. An important parameter for calculating the attenuation at low energies is the highest X-ray energy which can be emitted by the atom. This value is normally referred to as the k_α and κ_β spectral lines (Fig. 4.12). The energy values of both the X and the γ ray are above any k_α line of light elements which make living tissue. As a comparison, the k_α line of Calcium in the bones is at 3.7 keV. Therefore, the gamma ray attenuation in the body is very small. Most of the γ and X radiation escapes the patient body and can be detected outside. For each β^- particle emitted by the Iodine-131, there is a γ ray of a specific energy emitted by the Xe nucleus and an X-ray emitted by the Xe atom. By measuring the γ activity outside the patient body, we also measure the β activity of the administered dose. The X-ray activity for one particular energy transition can be calculated with atomic transition probabilities and *selection rules* and its calculation is outside the scope of this book.

5. *How long the patient must be in isolation to see its activity to decay to 500 Bq, neglecting any disposal by fluid exchanges?*

$$A(t) = (1/\tau)N(t) = (1/\tau)N_0 e^{-\frac{t}{\tau}} = A_0 e^{-\frac{t}{\tau}}$$

$$\frac{A(t)}{A_0} = e^{-\frac{t}{\tau}}$$

$$t = \tau \ln \frac{A_0}{A(t)} = 9.95 \times 10^5 \times \ln \frac{1.8 \times 10^9}{500} = 15.2 \times 10^6 \text{ s} = 174 \text{ days}$$

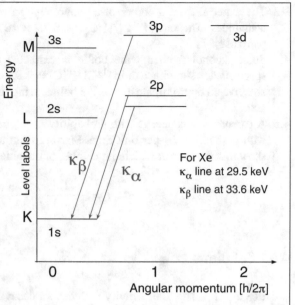

Fig. 4.12 The most intense sources of X-rays are atomic transitions to the lowest electron energy level. The energy levels are named as K, L, M or $1, 2, 3$, according to the main energy level and "s, p, d" are states of angular momentum $0, 1, 2$, respectively. Transitions from states, or *orbitals* $2p$ to $1s$, originate what are called k_α spectral lines, which are typically in the X-ray range. Transitions from $3p$ to $1s$ originate the k_β lines, which have higher energy than the k_α lines, but often have a lower intensity

6. *90% of the iodine is eliminated by the body in 3 days. Compare this time with the decay time above and suggest the appropriate measures to the hospital.* In reality, patients have to stay in isolation only a few days, because iodine is eliminated by the body much faster. However, patient's fluids have to be considered as radioactive waste for a few months.

4.7 Problems

4.1 One of the products of the decay chain of ^{241}Am is ^{213}Bi (Bismuth), which undergoes both α and β decay (see text). We have to measure the activity of a sample of pure ^{213}Bi. Our particle detector is wrapped in a thin aluminium foil and is only sensitive to beta particles, with an efficiency of 80%; our experimental set-up has a geometrical acceptance of 50%. At a given time, we measure a counting rate of 100 Hz. What is the total activity of the source? We measure the activity for 20 s, at regular intervals of 10 min, we plot the data, and we fit with an exponential curve, obtaining a mean decay time of 45.6 min. Is any correction needed to this measurement due to the fact that we only observe one decay mode? What value of the lifetime would we have measured if the detector were sensitive to α particles only?

4.2 Radioactive humans: an average human body of 70 kg contains 18% of carbon. With the data present in the text, calculate the "human activity" due to ^{14}C.

4.3 The material from a corner of a papyrus is analysed with mass spectrometer techniques. The ratio $[^{14}C]/[^{12}C] = 0.79 \times 10^{-12}$. What is the age of the papyrus?

4.4 Some alcohol from a wine bottle is extracted and analysed with a mass spectrometer, showing a ratio $[^{14}C]/[^{12}C] = 1.95 \times 10^{-12}$. Explain the anomaly and give an approximate age of the bottle, using the data which is given in the text.

4.5 A parent nucleus decays with probability λ_1 per unit time. Its daughter decays with probability λ_2 per unit time. Starting from a pure sample of the parent, show that the time at which the number of daughter nuclei is a maximum:

$$t_{max} = \frac{1}{\lambda_2 - \lambda_1} \ln\left(\frac{\lambda_2}{\lambda_1}\right).$$

4.8 Solutions

Solution to 4.1   The total activity includes all decays, in this case both α and β, and is counted over the whole solid angle. Therefore, we have to correct by the detector efficiency and by the solid angle *acceptance* and by the branching fraction:

$$A = \frac{\text{rate}}{\text{acceptance} \times \text{efficiency} \times \text{BR}} = \frac{100}{0.978 \times 0.8 \times 0.5} = 255.6 \text{ Bq}$$

In general, efficiency and acceptance have an experimental systematic error, which is added to the statistical error of the counting rate. No correction is needed to the measurement: the lifetime is a characteristic of the nuclide. When measuring the lifetime with a detector which is sensitive only to α particles, we would observe exactly the same value.

Solution to 4.2   The amount of carbon in a 67-kg human body is 12.05 kg, which is about 10^3 moles. We know that the concentration of $[^{14}C] = 1.30 \times 10^{-12}$, so the number of atoms of radiocarbon is

$$N_{14} = 10^3 \times N_A \times \left[^{14}C\right] = 10^3 \times 6.02 \times 10^{23} \times 1.30 \times 10^{-12} = 7.82 \times 10^{14}$$

Now, the average lifetime is

$$\tau = 8627 \text{ years} \times 31.536 \times 10^6 \text{ s/year} = 2.721 \times 10^{11} \text{ s}$$

The activity is

$$A = 1/\tau N_{14} = \frac{7.82 \times 10^{14}}{2.721 \times 10^{11}} = 2.87 \times 10^3 \text{ Bq} = 2.87 \text{ kBq}$$

In addition to a constant bombardment by cosmic rays, the human body is subject to radiation from inside. Also, radioactive K is a source of radiation.

Solution to 4.3   From Eq. (4.42):

$$\Delta t = -\tau \ln \frac{N(t)}{N_0} = -8627 \text{ years } \ln \frac{0.79 \times 10^{-12}}{1.30 \times 10^{-12}} = 4300 \text{ years before present}$$

The papyrus can be dated to 2300 BC.

Solution to 4.4   The radiocarbon concentration of the sample has a value $[^{14}C]/[^{12}C] = 1.95 \times 10^{-12}$, which is higher than the standard concentration, 1.30×10^{-12}. This means that the sample can be dated in a period during the so-called bomb-peak of ^{14}C. Neglecting the radioactive decay, which has a lifetime of 8627 years, we can use the exponential decay of the *excess* of radiocarbon, which has a lifetime $\tau = 23$ years, to find the year when the radiocarbon concentration was the same as the one found in the sample. We can use the same formulas, but for the excess of $[^{14}C]$, which reached its maximum value of twice the normal value in 1965.

$$N(t) - N_0 = (N_{\max} - N_0)e^{-\Delta t/\tau} \Rightarrow \Delta t = -\tau \ln((N(t) - N_0)/N_0) =$$

$$= 23 \times \ln((1.95 - 1.30)/1.30) = 16 \text{ years after the peak} \approx 1981$$

Solution to 4.5   The problem is simply solved by finding the maximum of the function in Eq. (4.36):

$$N_2(t) = N_1(0) \frac{\lambda_1}{\lambda_2 - \lambda_1} \left(e^{-\lambda_1 t} - e^{-\lambda_2 t}\right)$$

$$\frac{dN_2}{dt} = 0 = \frac{\lambda_1}{\lambda_2 - \lambda_1} \left(-\lambda_1 e^{-\lambda_1 t} + \lambda_2 e^{-\lambda_2 t}\right)$$

for $\lambda_1 \neq 0$

$$\lambda_1 e^{-\lambda_1 t} = \lambda_2 e^{-\lambda_2 t} \text{ and taking logs on both sides:}$$

$$-\lambda_1 t = \ln \frac{\lambda_2}{\lambda_1} (-\lambda_2 t)$$

$$(\lambda_1 - \lambda_2) t_{\max} = \ln \left(\frac{\lambda_2}{\lambda_1}\right)$$

Bibliography and Further Reading

H.D. Graven, Impact of fossil fuel emissions on atmospheric radiocarbon and various applications of radiocarbon over this century. Proc. Natl. Acad. Sci. USA **112**(31), 9542–9545 (2015)

B.R. Martin, *Nuclear and Particle Physics: An Introduction* (Wiley, Chichester, 2011)

E. Segré, *Nuclear and Particle Physics* (W. A. Benjamin, Reading, 1977)

Chapter 5
Passage of Radiation Through Matter

5.1 Introduction

In this chapter a more practical aspect of radiation is described: its interaction with matter. It has importance both for shielding and for detecting radiation. We'll limit the scope to X and γ rays for the electromagnetic radiation, to charged particles like α, $\beta^{\pm}$ and cosmic rays and, very schematically to *neutrinos*, leaving neutron interactions to the nuclear physics chapter.

Radiation interacts with matter by means of *scattering* processes: the initial radiation can be absorbed, deviated or can be transmitted (Fig. 5.1). In the first case all the initial energy is released in the medium, in the second case only a fraction of it. Initially we'll consider the common aspects of scattering, then the peculiarities of each type of radiation will be described. In the initial sections we focus on what happens to the radiation, while the effects on the material will be covered in the last two sections of this chapter: radiation detectors and biological effects.

5.2 The Effective Cross Section

We consider the case where we have a flux of incident radiation, which is measured in terms of number N_i of incident *photons* or charged particles per unit surface per second:

$$J = \frac{N_i}{S\,t},\qquad (5.1)$$

where S is the surface and t the time duration of the process. We assume that within that surface S the density is constant, or in other words J is constant in the considered time interval. We define the *intensity* I as the number of incident

© Springer Nature Switzerland AG 2018
S. D'Auria, *Introduction to Nuclear and Particle Physics*,
Undergraduate Lecture Notes in Physics,
https://doi.org/10.1007/978-3-319-93855-4_5

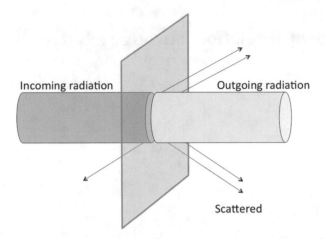

Fig. 5.1 Schematics of passage of radiation through a thin layer of material. Part of the radiation is transmitted, part is scattered or absorbed

particles per unit of time:

$$I = \frac{N_i}{t} \; ;$$

(5.2)

In case the incoming radiation beam is not uniformly distributed, the intensity is the integral of the flux over the surface:

$$I = \int_S J(x, y) dx \, dy$$

(5.3)

The material is characterised by its density and by its chemical composition. It is reasonable to expect that a layer of liquid nitrogen absorbs more radiation than a layer of air, which is mostly nitrogen, of the same thickness, just because there are more scattering centres per unit of volume. Each scattering centre can be more or less efficient in scattering, depending on its apparent "size". This is parametrised by another quantity, which depends both on the impinging radiation and on the material: it is the *effective cross* section, it is indicated with σ and is a characteristic property of all scattering processes. It depends on the nature and kinetic energy of the incoming particle, on the process that takes place and, of course, on the material composition. The concept of cross section can be extended to other situations, like colliding particle beams. From a qualitative point of view it is reasonable to expect that a radiation with a small wavelength or high energy (Eq. (3.3)) is less attenuated than a radiation with larger wavelength, for a given material. For a given wavelength it is reasonable to expect that atoms with larger numbers of electrons each have a larger probability to interact with radiation compared with a material with lower Z. The term *cross section* derives from experiments on fixed targets, like the scattering of α particles on a gold foil, as first done by E. Rutherford, H. Geiger and E. Marsden

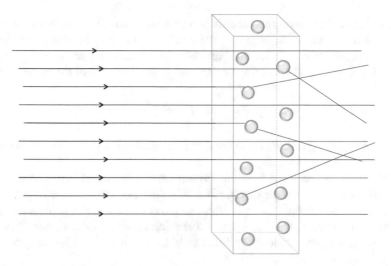

Fig. 5.2 Scattering of particles from a target. The scattering probability is larger for a larger density of the target and depends on the quantum probability of interacting with a single scattering centre. This quantity is the *cross section* σ, the apparent size of a scattering centre, which depends on the type and energy of the incident radiation. Figure adapted from Povh et al. (2015)

Ernest Rutherford (1871–1937) from New Zealand was awarded the Nobel prize in chemistry in 1908 for having demonstrated that radioactive decays induce a transmutation of elements, but he is most famous for the experiment that demonstrated that majority of atomic mass is concentrated in the nucleus.

in 1909, which demonstrated the existence of a relatively massive nucleus inside Au atoms. The radiation may interact with atoms, single electrons or nuclei in the target, which appear to have an *effective* size, which changes as a function of the energy and the type of the incoming particle. The probability of hitting a target is proportional to the target area perpendicular to the direction of the projectiles, as sketched in Fig. 5.2. When there are many targets, the probability also depends on their density, in terms of targets per unit of surface. In general, we know the density ρ of a material in terms of kg/m^3 or g/cm^3; we obtain the *atomic density* $\mathcal{N}$ by dividing ρ by the mass of a single atom. Equivalently, if M indicates the atomic mass of the material, which we assume is made by a pure element, like gold or lead, and N_A represents Avogadro's number,

$$\mathcal{N} = \frac{\rho\, N_A}{M} \tag{5.4}$$

The number of scattering centres "seen" by an incoming particle depends on this density and on the thickness δx of the absorber. In the following we assume that the thickness is small so that we have no shadowing effect, and that the material is not a crystal, so that there is no special direction of propagation. The number of scattered or absorbed particles, N_s, is proportional to the number of incident particles N_i, to the cross section $\sigma_p(E)$ for that particular process and for that particular energy range, to the density of scattering centres and to the thickness of the target:

$$N_s = N_i \frac{\rho N_A}{M} \delta x \, \sigma_p(E) \quad \text{and} \quad R_s = I_i \frac{\rho N_A}{M} \delta x \, \sigma_p(E) \,, \tag{5.5}$$

where the scattering rate R_s is the number of scattering interactions per unit of time and I_i is the incident intensity.

Conversely, we can read from the formulas above that the effective cross section of a scattering process is the number of interactions per unit time per target particle per unit of incident flux. So, the effective cross section is the expression of the quantum-mechanical probability that a scattering occurs. As the scattering process is the same for any boost along the direction of the incoming particles, its description is relativistically invariant.

Given a type of radiation and a material, more than one process may be possible. In this case, the total cross section is the sum of the cross sections of the single processes. As an example, looking at Fig. 5.6, for photons of 1.5 MeV all three processes are possible: Compton scattering, photoelectric effect and pair creation. The total cross section, which we use to calculate the attenuation coefficient of the radiation, is the sum of these three cross sections:

$$\sigma_{\text{tot}} = Z\sigma_C + \sigma_{\text{pe}} + \sigma_{ee} \tag{5.6}$$

If the target material is a chemical compound or a mixture, the individual scattering rates of each element are added.

The total cross section has physical dimensions of $[L^2]$; in nuclear and particle physics the unit is the *barn*:

$$1 \text{ barn} = 10^{-28} \text{ m}^2 = 100 \text{ fm}^2 \tag{5.7}$$

and its sub-multiples millibarn (mb), nanobarn (nb), picobarn (pb) and femtobarn (fb).

For a given scattering process it may be relevant to calculate also the probability that the scattering occurs to a particular final state, e.g. the probability (or rate) to scatter a photon to a given solid angle interval around a direction, or in a given energy or momentum interval. The formulas are the same as in Eq. (5.5) but we use the *differential cross section*, which are indicated as (Fig. 5.3):

$$\frac{d\sigma}{d\Omega}; \quad \frac{d\sigma}{dp}; \quad \frac{d\sigma}{dE} \tag{5.8}$$

for solid angle, momentum and energy, respectively. These distribution functions give the scattering probability to a given interval of final states. Double differential

Fig. 5.3 $d\Omega$ is the solid angle. The total cross section for a process is proportional to the interaction probability. The differential cross section $d\sigma/d\Omega$ gives the probability that a given outgoing particle is scattered at a certain solid angle. Another example is $d\sigma/dE$ for the probability of an outgoing particle to have an energy between E and $E + dE$

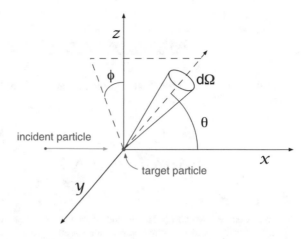

cross sections are two-dimensional distributions of scattering probability:

$$\frac{d^2\sigma}{d\Omega\,dp} \tag{5.9}$$

Integrating over the whole solid angle and over the momentum, we recover the cross section.

5.3   Scattering of Electromagnetic Radiation

Electromagnetic waves with short wavelength, like X- and γ-rays, interact with matter in four principal processes:

- Rayleigh scattering occurs when light interacts with an atom and is scattered at a different angle, without ionising or exciting it.
- The photoelectric effect occurs when light interacts with an atom and expels an electron (Fig. 5.4).
- Compton scattering occurs when a photon interacts with a single electron of an atom (Fig. 5.5).
- The photon can "split" into an electron and positron pair if it has enough energy and interacts with the electric field of a nucleus. This process is called *electron–positron pair creation*.

For a given photon energy, each of the processes above has a different probability of occurring. The probability of each process depends on the photon energy in a different way. The total *cross section* is the sum of the cross section, or Q-M. probability, of each process, as shown in Fig. 5.6. In the Rayleigh scattering process the electromagnetic radiation interacts with the atom or molecule as a whole and is just deviated, not absorbed. For a given atom or molecule, the cross section varies with the wavelength as $1/\lambda^4$, it is larger for short wavelength, like those corresponding to the blue colour; this qualitatively explains the blue colour of the sky.

Fig. 5.4 Photoelectric effect. Note that the photon is absorbed by the atom, which expels an electron

Fig. 5.5 Compton scattering. The photon interacts with the electron, transferring momentum. The electron is emitted and also a photon is emitted at an angle with respect to the impinging photon

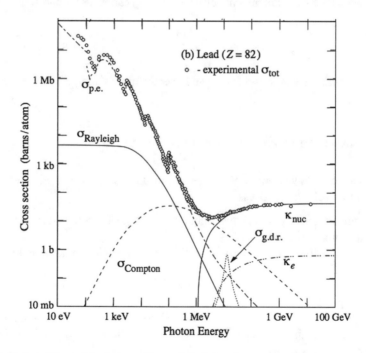

Fig. 5.6 Cross section of lead atoms Pb, as "seen" by photons in a large energy range. The calculated cross sections of single processes are also shown. The absorption and scattering of photons in the matter is the sum of several fundamental processes. Reprinted with permission from M. Tanabashi et al. (Review of Particle Physics), Phys. Rev. D, vol. 98-1, p. 454 (2018). Copyright (2018) by the American Physical Society. The unit used in the cross section is the *barn*, (b), defined as $1\,b = 100\,fm^2 = 10^{-28}\,m^2$

5.4 Attenuation of Electromagnetic Radiation

If the energy of the light is larger than the minimum ionisation energy of the target atoms the *photoelectric effect* may occur. In this case the atom absorbs a photon and expels an electron. With higher photon energies electrons from inner shells can be excited. Qualitatively, inner electron shells are also closer to the nucleus, making a smaller target for the photon; also the wavelength of the photon is smaller, so both effects contribute to a decrease of the cross section when increasing the photon energy.

The practical limits in terms of photon energy for the photoelectric effect to occur for a given element are given by the lowest ionisation energy and by the energy corresponding to the k_β X-ray emission line of the atom, which is close to the maximum energy transition in an atom, as shown in Fig. 4.12. For photon energies above the *K-edge* the photoelectric effect still occurs, but at a lower rate.

The photoelectric process completely removes photons from the initial beam. Starting from Eq. (5.5), the scattering rate R_s is the amount of intensity removed from the beam:

$$R_s = I_{\text{incident}} - I_{\text{transmitted}} = -\delta I = \rho \sigma \frac{N_A}{M} I_i \, \delta x \ . \tag{5.10}$$

Using

$$\mu = \rho \frac{N_A}{M} \sigma = \mathcal{N} \sigma \ \text{we have:} \ \delta I = -\mu \, I_i \, \delta x; \quad \frac{\delta I}{\delta x} = -I_0 \, \mu \tag{5.11}$$

Replacing small quantities with infinitesimal quantities, we have a differential equation

$$\frac{dI(x)}{dx} = -\mu \, I(x) \ , \text{whose solution is:} \ I(x) = I_i \, e^{-\mu x} \tag{5.12}$$

This function is shown in Fig. 5.7. The parameter μ has physical dimensions $[L^{-1}]$ and is called the *attenuation coefficient*. It depends on the density of the absorber and on the cross section, which is a function of the energy of the radiation. In general a given material may have different densities, so the density is factored out in most tables and calculations which report μ/ρ in cm^2/g. This means that we need to multiply this value by the density of the material to obtain the value to use in Eq. (5.12).

The cross section for the photoelectric effect, in a given material, varies with the photon energy as

$$\sigma_{\text{p.e.}} \propto \frac{1}{E_\gamma^3} \tag{5.13}$$

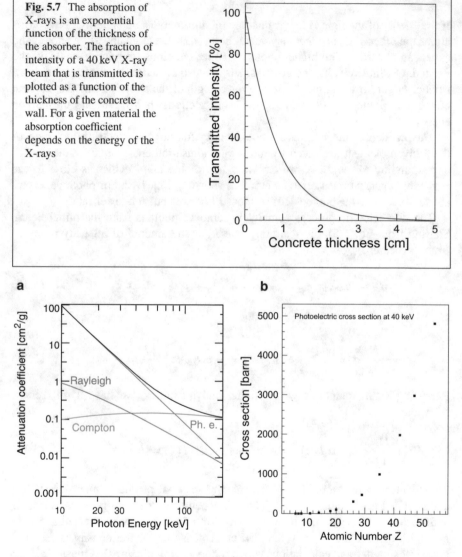

Fig. 5.7 The absorption of X-rays is an exponential function of the thickness of the absorber. The fraction of intensity of a 40 keV X-ray beam that is transmitted is plotted as a function of the thickness of the concrete wall. For a given material the absorption coefficient depends on the energy of the X-rays

Fig. 5.8 (**a**) Left: attenuation coefficient of calcium as a function of the photon energy. The contributions from three processes (Rayleigh, Compton scattering and photoelectric) are shown separately, and as a sum.
(**b**) Right: cross section per atom for the photoelectric process for 40 keV X-ray photons, as a function of the atomic number of the absorber. The best fit to these data gives a power law with exponent 4.2 at this energy. Data from National Institute of Standards and Technology Standard Reference Data Program

but the curve presents several discontinuities at values corresponding to the energy levels of the atoms, as shown in Fig. 5.6. For a given photon energy, the photoelectric cross section increases with the fourth power of the atomic number Z, as shown in Fig. 5.8b:

$$\sigma_{\text{p.e.}} \propto Z^n \; ; \; \text{with} \quad 4 \leq n \leq 4.6 \, , \quad \text{at fixed } E_\gamma \tag{5.14}$$

Below a threshold energy, which depends on the material, photons are unable to ionise and can only be diffused by the material. Above a material-dependent energy value, photons interact predominantly with electrons rather than with the atom as a whole.

5.5 Compton Scattering

Compton[1] scattering occurs when photons interact with single electrons in the cloud of an atom. The cross section is a function of the photon energy and increases linearly with the electron density, which is proportional to Z. From the practical point of view it becomes the dominant scattering mechanism for photons in an energy range around $0.1–1$ MeV, the limits depending on the material Z.

The kinematics are simple: a massless particle hits a massive particle, an electron, which is considered to be at rest. With the help of special relativity we can calculate the energy of the scattered photon as a function of the scattering angle and of the initial photon energy.

For a particle with $m \neq 0$ and 4-momentum $\mathbf{p} = (\frac{\gamma mc^2}{c}, p_x, p_y, p_z)$ we calculate the invariant mass $\sqrt{\mathbf{pp}}$ in the reference frame where the particle is at rest and $\vec{p} = 0$ and $\gamma = 1$. As it is a 4-vector, this quantity, the invariant mass, remains invariant, and is $\sqrt{\mathbf{pp}} = mc$, or in a general form

$$\mathbf{pp} = \frac{E^2}{c^2} - p^2 = m^2 c^2 \quad \text{or} \tag{5.15}$$

$$E^2 - p^2 c^2 = m^2 c^4 \tag{5.16}$$

It is invariant because if we consider the entire system, before and after a decay or any reaction, it does not change.

The scattering process we are going to describe is:

$$\gamma + e^- \rightarrow \gamma + e^- \tag{5.17}$$

We neglect the binding energy of the electron to the atom, because this is typically of the order of few eV, and solve the kinematic in the reference frame where

---

[1]After Arthur H. Compton (1892–1962) from USA. He was awarded the Nobel prize in 1927 "for the discovery of the effect named after him".

Fig. 5.9 Compton scattering

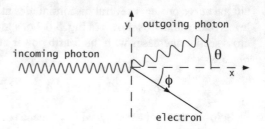

the electron is initially at rest. It is clear that the scattering process occurs in a
geometrical plane containing the momenta of all particles involved. We assume
this is the $x - y$ plane, and we neglect the z coordinate. In general, the incoming
and outgoing photons will not have the same energy. Let E_0 be the energy of the
incoming photon and E_1 the energy of the outgoing photon, as shown in Fig. 5.9.
We apply momentum and energy conservation independently, and use the relativistic
formulae:

$$p_x : \qquad\qquad E_0/c = (E_1/c) \cos \theta + p \cos \phi; \qquad\qquad (5.18)$$

$$p_y : \qquad\qquad (E_1/c) \sin \theta - p \sin \phi = 0; \qquad\qquad (5.19)$$

$$E : \qquad\qquad E_0 + mc^2 = E_1 + E_e \qquad\qquad (5.20)$$

where $p = |\vec{p}|$ is the electron momentum, m is the electron mass and the angles ϕ
and θ are measured between outgoing particles and the direction of the incoming
photon. E_e is the electron energy.

From momentum conservation we have:

$$E_1 \sin \theta = pc \sin \phi; \qquad\qquad (5.21)$$

$$E_0 - E_1 \cos \theta = pc \cos \phi; \qquad\qquad (5.22)$$

We square and sum:

$$E_1^2 \sin^2 \theta + (E_0 - E_1 \cos \theta)^2 = p^2 c^2 (\sin^2 \phi + \cos^2 \phi); \qquad\qquad (5.23)$$

$$E_1^2 \sin^2 \theta + E_0^2 + E_1^2 \cos^2 \theta - 2E_0 E_1 \cos \theta = p^2 c^2; \qquad\qquad (5.24)$$

$$E_1^2 + E_0^2 - 2E_0 E_1 \cos \theta = p^2 c^2; \qquad\qquad (5.25)$$

now $p^2 c^2 = E_e^2 - m^2 c^4$ from the electron 4-momentum. Using energy conservation
we have

$$E_e^2 = ((E_0 - E_1) + mc^2)^2; \text{ subst. in 4-momentum} \qquad\qquad (5.26)$$

$$p^2 c^2 = (E_0 - E_1)^2 + m^2 c^4 + 2mc^2 (E_0 - E_1) - m^2 c^4; \qquad\qquad (5.27)$$

$$p^2 c^2 = (E_0 - E_1)^2 + 2mc^2(E_0 - E_1) = E_0^2 + E_1^2 - 2E_0 E_1 \cos \theta; \qquad (5.28)$$

$$E_0^2 + E_1^2 - 2E_0 E_1 + 2mc^2(E_0 - E_1) = E_0^2 + E_1^2 - 2E_0 E_1 \cos \theta; \qquad (5.29)$$

$$E_0 - E_1 = \frac{E_0 E_1}{mc^2}(1 - \cos \theta) \qquad (5.30)$$

Passing from energies to wavelengths: $E_j = h\nu_j = hc/\lambda_j$

$$E_0 - E_1 = hc\left(\frac{1}{\lambda_0} - \frac{1}{\lambda_1}\right) = hc\left(\frac{\lambda_1 - \lambda_0}{\lambda_1 \lambda_0}\right) \qquad (5.31)$$

$$\lambda_1 - \lambda_0 = \frac{hc}{mc^2}(1 - \cos \theta); \qquad (5.32)$$

The constant quantity

$$\lambda_e = \frac{hc}{m_e c^2} = 2.4263102367(11) \times 10^{12} \text{ m} \qquad (5.33)$$

is called the *Compton wavelength of the electron*, where m_e is the electron mass and h is Planck's constant. The two digits in parenthesis represent the experimental error on the previous two digits. So far we have only used relativistic kinematics: nothing is said about the *interaction* between the photon and the electron: the above discussion is valid no matter what interaction is involved. The interaction enters into play when we calculate the cross section, i.e. the probability of the Compton scattering. This was done by Klein and Nishina in 1928.

5.6 The Cross Section for Compton Scattering

Klein[2] and Nishina[3] in 1928 calculated for the first time the Compton cross section using quantum mechanics.

$$\frac{d\sigma}{d\Omega} = \frac{1}{2}\alpha^2 r_c^2 P^2(E_\gamma, \theta)\left[P(E_\gamma, \theta) + \frac{1}{P(E_\gamma, \theta)} - 1 + \cos^2 \theta\right] \qquad (5.34)$$

where $d\Omega = \sin \theta \, d\theta \, d\phi$ and

$$P(E_\gamma, \theta) = \frac{1}{1 + (E_\gamma/m_e c^2)(1 - \cos \theta)} \quad \text{and} \qquad (5.35)$$

---

[2] Oskar Klein (1894–1977), from Sweden. He's also known for the Kaluza–Klein theories, which speculate the existence of additional space dimensions.
[3] Yoshio Nishina (1890–1951) from Japan. He also discovered the isotope ^{237}U.

$$r_c = \frac{\hbar c}{(2m_e c^2)} \qquad (5.36)$$

is the "reduced" Compton wavelength of the electron, $(\hbar = h/2\pi)$ is the "reduced" Planck's constant and

$$\alpha = \frac{e^2}{4\pi\epsilon_0 \hbar c} \approx 1/137 = 7.297 \times 10^{-3}, \qquad (5.37)$$

We call $\alpha = \alpha_{EM}$ the *fine structure constant* for the electromagnetic interaction. It depends on the electron charge e, it is a dimension-less number, which gives the "strength" of the electromagnetic interaction, and its value is totally independent of the unit system we use.

$\epsilon_0 = 8.854 \times 10^{-12}$ F/m is the permittivity of vacuum.

Note how the Klein and Nishina formula for the differential cross section of the Compton scattering *depends on the fourth power of the electric charge*, while the electric charge does not appear in the kinematic formulas of Compton scattering.

The K-N formula gives the angular dependence of the cross section: it gives the probability that a photon is scattered at a given angle. As we have seen already for the kinematic of the scattering, the process only depends on one angle: the one between the initial and the final direction of the photon. We say that there is a *cylindrical symmetry* around the axis of the initial photon direction. In other words, the probability of scattering along the azimuth angle ϕ is uniform. Note also that the cross section decreases when the photon energy E_γ increases. The cross section is referred to each scattering centre, i.e. electrons; so we need to multiply it by Z to obtain the Compton cross section for a material, as in Eq. (5.6).

5.7 Heavy Charged Particles

In this case the mass of the "projectile" particle is much larger than the mass of the "target" particle in the material, which is mostly electrons. This means we are dealing with protons, alpha particles or muons, or other particles which we'll meet later.

The incoming charged particle loses part of its kinetic energy in a series of many interactions. In each of them the projectile transmits a small part of its kinetic energy to the electrons in the target. The incoming particle is also deviated slightly from its initial direction. The term *multiple scattering* is used to indicate this effect. The Bethe-Bloch formula, below, describes the energy loss per unit of length of a heavy charged particle when colliding with the electrons in a medium:

$$-\left\langle \frac{dE}{dx} \right\rangle = \left(\frac{e^4}{4\pi\epsilon_0^2} \right) \frac{z^2}{m_e v^2} (Z\mathcal{N}) \ln\left(\frac{2m_e v^2}{I} \right) \qquad (5.38)$$

This is the non-relativistic version of the formula, which is valid for a kinetic energy of the particle in the interval from 0.5 to 500 MeV, when dealing with protons, alpha particles or *mesons*. This formula is calculated in many, and more advanced, textbooks (Cerrito 2017; Fernow 1986; Segré 1977). We can use it as a starting point to highlight the features of the passage of electrically charged particles.

The electron electric charge e is present in the formula, because the energy loss is due to electromagnetic interaction and, once again, it enters to the fourth power, just like in Compton scattering cross section. The electric charge of the projectile z (in units of e) enters the formula quadratically, so an alpha particle loses 4 times its energy with respect to a proton of the same velocity v; this also enters the formula quadratically and is measured with respect to the target, which is at rest. Z indicates the atomic number of the target, or the number of electrons per atom and enters linearly in the formula, like the atomic density of the material (Eq. (5.4)), which is indicated with $\mathcal{N}$. In other words the term $(Z\mathcal{N})$ is the average electron density. The parameter I is the average ionisation energy of the target, and is shown in Fig. 5.10

Felix Bloch (1905–1983) from Switzerland was awarded the Nobel prize in 1952 for his work on precision measurements of nuclear magnetic moments. He was the first director general of the European laboratory CERN.

as a function of Z. These collisions occur between the charged particle and the electrons in the target atoms and therefore the electron mass m_e is the only mass

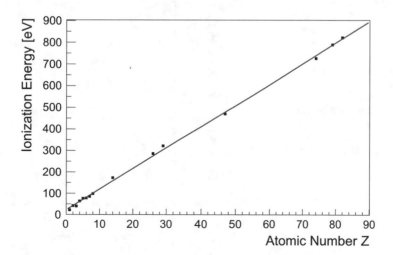

Fig. 5.10 Average ionisation energy as a function of the atomic number. The line represents a fit to the data: $I \approx 22.8 + 9.7Z$

appearing explicitly in the formula. In general also in this case the density of the
material is factored out and $1/\rho \langle dE/dx \rangle$ is called the *stopping power*. The mass of
the projectile particles, m_p, appears if we re-write Eq. (5.38) in terms of the kinetic
energy $E = 1/2 m_p v^2$ of the incoming particle:

$$-\frac{1}{\rho}\left\langle \frac{dE}{dx} \right\rangle = K \frac{m_p}{m_e} \frac{1}{E} \frac{z^2 Z}{M} \ln\left(\frac{4m_e E}{m_p I}\right) \tag{5.39}$$

where

$$K = 2\pi \, \alpha^2 \hbar^2 c^2 N_A = 0.078 \text{ MeV}^2 \text{ cm}^2$$

It is clear that at constant density, materials with higher Z have a higher stopping power. For higher energies, the Bethe-Bloch formula has to be corrected
by a relativistic factor γ^2 inside the logarithmic term, which gives rise to an
increase of energy loss with the particle energy. The energy loss is minimum for
charged particles in the interval 0.1–1 GeV, which are called *minimum ionising
particles*.

At even higher energies radiative losses become important, with a mechanism
which is very important for electrons. Another clear limitation of Eq. (5.39) is
evident in the low kinetic energy limit, where the logarithmic term would become
zero and then negative, which is unphysical. Therefore, there is an intrinsic cut-off,
which depends on the ionisation energy. Equation (5.39) which is plotted in Fig. 5.11
for muons in copper and Eq. (5.38) have to be used with care. The stopping power
of copper for *muons* is shown in Fig. 5.12, for a large energy range. In this figure all
effects and corrections are taken into account.

Fig. 5.11 The simplified
form of the Bethe-Bloch
formula, Eq. (5.39), can be
used to calculate the energy
loss of muons in copper,
within its interval of validity:
as long as the particle is
non-relativistic and its kinetic
energy is not too low. The
results of the calculation of
the muon stopping power for
copper is shown as a function
of the muon kinetic energy

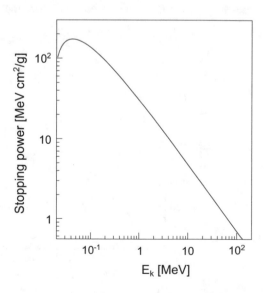

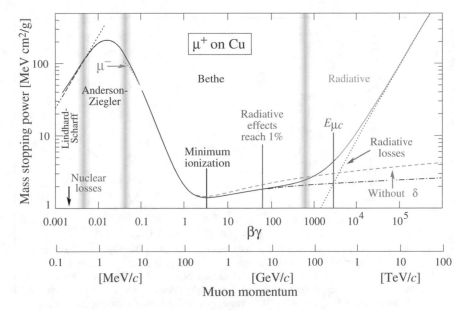

Fig. 5.12 Specific energy loss in copper for muons as a function of their initial momentum. Reprinted with permission from M. Tanabashi et al. (Review of Particle Physics), Phys. Rev. D, vol. 98-1, p. 447 (2018). Copyright (2018) by the American Physical Society. Muons have a mass about 200 times larger than electrons and are electrically charged. To obtain the energy loss per unit of length the values on the y axis have to be multiplied by the copper density

In case of charged particles traversing a material, their flux remains constant, but the kinetic energy of each particle decreases. This is very different from the case of photons, whose number decreases exponentially inside the material. In case of thick absorbers, particles can lose all their energy inside the material and come to a rest after a well-defined *range R* which depends on their initial kinetic energy E_k. To calculate the range we need to integrate Eq. (5.38) from E_k to zero kinetic energy. The logarithmic term makes the analytic integration difficult, but above a certain initial energy it just adds a constant. If we consider the logarithmic term as a constant we have:

$$R = \int_{E_k}^{0} dE \frac{1}{\langle -dE/dx \rangle} \approx \text{Const} \int_{E_k}^{0} E \, dE = {}^{1}\!/_{2}\,\text{Const}\,E_k^2 \qquad (5.40)$$

In the energy range where the average energy loss varies with the kinetic energy as $1/E$, the range of the particle increases quadratically with its energy. Writing explicitly the entire formula, and using the log of the average energy, from Eq. (5.39) we obtain:

$$R \approx \frac{1}{2K}\frac{m_e}{m_p}\frac{M}{Z\rho}\frac{1}{z^2}\left[\ln\left(\frac{2m_e E_k}{m_p I}\right)\right]^{-1} E_k^2 \; ; \qquad (5.41)$$

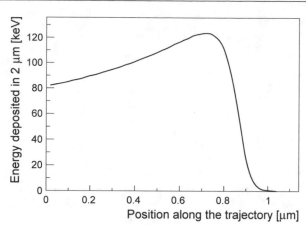

Fig. 5.13 Energy deposition as a function of the position along the trajectory for 5 MeV
α-particles in copper. Most of the kinetic energy is deposited in a narrow layer near the end
of the particle trajectory. This feature is used in hadron therapy, where the peak is much more
pronounced, and in medical applications of α-emitting radionuclides. The curve is named
after William Henry Bragg (1862–1942), from England, he taught physics in Adelaide,
Australia, for 20 years, then in Leeds and London. He was awarded the Nobel prize in 1915
with his son Lawrence for their research on X-ray diffraction in crystals

the range is inversely proportional to the density of the material, as expected. This
formula has its own limitations: for instance, it calculates the path length, but the
trajectory is not a straight line, so we don't obtain the depth in the material where
the particle would stop. However, it can be used to calculate the upper limit of
the range of α particles in air or β particles in Al. For practical applications, like
doping of semiconductor materials by ion implantation, more refined calculations
are implemented in computer-based simulations.

The energy loss per length of the path is larger as particles progressively lose
their energy, as is evident from the $1/E$ term in Eq. (5.39). Therefore, most of
their energy is lost at the end of the path, where their kinetic energy is lowest.
Plotting the specific energy loss as a function of the penetration depth we obtain the
characteristic *Bragg curve*, which has a peak of energy deposition near the end of the
particle trajectories, as shown in Fig. 5.13. This feature is used in hadron therapy,
where a beam of protons is aimed at a cancer and its energy is tuned to have the
protons stop inside the cancer volume, where most of its energy is delivered. For
the same reason α-emitting radionuclides for medical treatment are more effective
in delivering high dose to cancer cells than the γ-emitting nuclides.

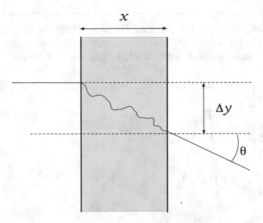

Fig. 5.14 Multiple scattering of charged particles traversing a material: they are deviated from their initial direction by an angle θ which depends on the material and on its thickness x

5.8   Charged Particles Traversing Thin Layers

Electrically charged particles lose kinetic energy by a large number of electron scattering events. After traversing a thin section of material the charged particles emerge with an angle θ with respect to the original direction (Fig. 5.14). The distribution of these angles is approximated by a Gaussian, and its standard deviation is inversely proportional to the particle momentum and proportional to the square root of the material thickness.

The energy loss distribution in thin layers, however, does not follow a Gaussian distribution. The reason is that occasionally scattering events occur with a large energy loss. The resulting distribution is skewed, with the average energy loss being considerably larger than the most probable energy loss. The average energy loss is still what calculated with Eq. (5.38), but the most probable energy loss is a more significant parameter. The distribution is described by the following integral function:

$$\phi(x) = \frac{1}{\pi} \int_0^\infty dt \, \exp(-t \log(t) - xt) \sin(\pi t) \qquad (5.42)$$

This is the so-called Landau distribution; it is a slightly complicated function: the variable x representing the energy loss is inside the integral and the function has no free parameter. The shape of the distribution is shown in Fig. 5.15. It reproduces well the energy loss in thin materials, as long as they are not extremely thin. In these cases the Landau distribution is considerably wider than the experimental distribution and better distributions have width and most probable value as parameters. The energy loss distribution is also called *straggling distribution*.

The distribution is named after Lev Davidovich Landau (1908–1968) from Russia. He was awarded the Nobel prize in 1962 for his work on superconductivity. He is also known for his textbook on theoretical physics, in ten volumes, written with Evgeny Lifshitz.

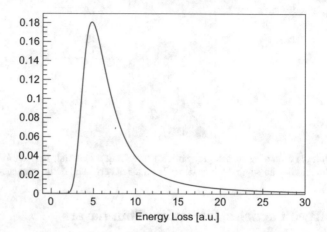

Fig. 5.15 The Landau distribution describes the energy loss of charged particles in thin absorbers, where they lose only a small fraction of their kinetic energy. Its analytical form is shown in Eq. (5.42). For very thin absorbers this formula slightly underestimates the width

5.9 Electron Energy Loss

Electrons have the peculiarity that are identical to the main target they interact with in the matter: other electrons.

Because of their small mass, electrons are subject to a large deceleration and emit a considerable electromagnetic radiation, which is called bremsstrahlung, German for "braking radiation". When they interact with other electrons both particles emit braking radiation, which interferes destructively at large distances. However, when they scatter off the electric field of atomic nuclei their acceleration emits electromagnetic radiation. The energy loss per unit of length, by radiation only, is given by the following formula:

$$-\left(\left(\frac{dE}{dx}\right)\right)_{rad} \approx \frac{4\mathcal{N}Z^2\alpha^3(\hbar c)^2}{m_e^2 c^4} E \ln \frac{183}{Z^{1/3}} \qquad (5.43)$$

where $\mathcal{N} = \rho N_A/M$ and N_A is the Avogadro's number, $\alpha \approx 1/137$, A and Z are the mass and atomic number of the target, E is the energy of the electron and $\hbar = h/2\pi$ and m_e is the electron mass. As above, we can just note some features, e.g. that this cross section increases linearly with the electron kinetic energy E.

We define the *critical energy of an absorber* E_c as the electron energy above which the energy loss by radiation is larger than the energy loss by ionisation. It

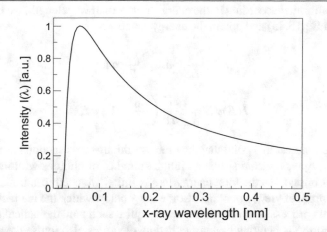

Fig. 5.16 The wavelength λ of the X-rays emitted by bremsstrahlung has a continuous distribution. The wavelength distribution of X-rays emitted by 40 keV electrons is calculated with Eq. (5.45) and shown here

varies as $E_c \propto Z^{-1}$: for solid material an approximate function is

$$E_c \approx \frac{610}{Z + 1.24} \text{ MeV} \tag{5.44}$$

and its values are in the range 5–350 MeV. Low-energy electrons, such as those produced by radioactive sources or used in cathode ray devices, lose most of their energy by ionisation; however, braking radiation is never completely negligible.

The braking radiation is emitted in the X-ray range: energetic electrons, when interacting with the matter, transform part of their kinetic energy into electromagnetic radiation. This mechanism occurs in X-ray tube generators. The emitted X-ray spectrum depends on the electron initial energy E_0, but has a universal shape, which is given by the Kramers' law:

$$I(\lambda)\, d\lambda = k \frac{iZ}{\lambda^2} \left(\frac{\lambda}{\lambda_{\min}} - 1 \right) d\lambda; \quad \text{where } \lambda_{\min} = \frac{hc}{E_0} \tag{5.45}$$

and $I(\lambda)\, d\lambda$ is the X-ray intensity at wavelength λ, i is the electron intensity. The shape of the wavelength spectrum is shown in Fig. 5.16.

At high wavelengths, the emission of X-rays is collinear with the electrons, and is absorbed by the same material and by any other material which may be present. This changes dramatically the spectrum shape at high wavelengths. At low electron energy ($E_0 < m_e c^2$) the emission of X-rays is maximum in the direction perpendicular to the electron beam and is polarised with the electric field oscillating along the electron direction. At higher energies ($E_0 \gg m_e c^2$) the radiation is emitted mostly along the initial direction of the electrons. As X-ray detectors are normally calibrated according to the energy, rather than wavelength, we can change variables in Eq. (5.45) and obtain the energy spectrum

$$\lambda = \frac{hc}{E}; \quad d\lambda = \frac{hc}{E^2} dE$$

$$I(E)dE = k\frac{iZ}{hc}\left(\frac{E_0}{E} - 1\right)dE \tag{5.46}$$

It should be noted at this point that the bremsstrahlung cross section has a so-called *infrared divergence*, meaning that an infinite number of photons with infinitesimal energy are emitted. As the total energy loss is finite, we need to introduce, from a theoretical point of view, a photon cut-off energy below which the interaction can be ignored. This process is not the only one to require such a mathematical treatment.

In addition to generating braking radiation, electrons also ionise the atoms, just like other charged particles. Electrons from the outer shells make a transition and occupy the inner shells or energy levels which have been left empty by the ionisation process, emitting electromagnetic energy, i.e. photons. The photon energy depends on the difference between the energy levels of the atoms. Therefore the material emits also a characteristic X-ray radiation, which depends on the material which is irradiated. The resulting spectrum is shown in Fig. 5.17, where the following effects are calculated:

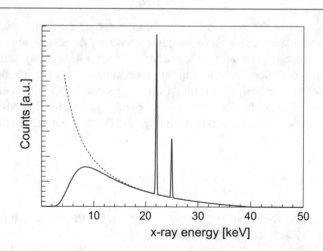

Fig. 5.17 Calculated energy spectrum of X-rays emitted by an X-ray tube with silver anode operated with 40 keV electrons. The bremsstrahlung spectrum is calculated with Eq. (5.46). The low-energy X-rays are absorbed by the anode itself and by the glass window; the absorption cross section is assumed as $\sigma_{pe} \propto 1/E_\gamma^3$. The shape of the un-filtered spectrum is shown as a dashed curve. The silver K_α and K_β lines are added as narrow Gaussian distributions. Compare it with the experimental spectrum shown in Fig. 5.23

- Bremsstrahlung emission, as in Eq. (5.46),
- Absorption of low-energy X-rays by the material itself and by the glass window,
- Emission of K lines by the material, which we assume to be silver.

5.10 Neutrino Interactions with Matter

Neutrinos will be introduced more formally in the next chapter, but we have already encountered them in describing the β decay. They are electrically neutral particles and their mass is not precisely measured, it is extremely small, but not zero. The present limit for all three types of neutrinos is at about $m_\nu < 1$ eV. They only interact with matter with the weak interaction. A feature of this interaction is that the cross section increases with the neutrino energy, up to very high energies. Nevertheless, the cross sections for various processes involving neutrinos range from 10^{-28} to 10^{-17} barn for energies in the range from 100 eV to 1 GeV. As a comparison, the Compton cross section in lead ($Z = 82$) is of the order of 10 barn (Fig. 5.6), which makes the cross section per electron 16 orders of magnitude larger. The energy loss mechanisms of neutrinos would be similar, in principle, to the energy loss of charged particles, but the multiple scattering occurs on a much larger scale, due to the extremely small cross section.

We can look back at the equation describing the photon attenuation, Eq. (5.11). Independently from the fate of the impinging particle, it describes the distribution of the depth of its first interaction with the matter; the parameter $1/\mu$ is the mean path before interaction. Using the largest value of neutrino scattering cross section, for a kinetic energy in the ≈ 1 GeV range (10^{-17} barn), in Eq. (5.11) the neutrino mean free path in iron is:

$$\mu_\nu = \rho \frac{N_A}{M} \sigma \; ;$$

$$\mu_\nu(Fe) \approx 7.87 \frac{g}{cm^3} \times \frac{6.022 \times 10^{23} \, mol^{-1}}{55.84 \, g \, mol^{-1}} \times 10^{-17} \times 10^{-24} \, cm^2$$

$$\mu_\nu(Fe) \approx 8.5 \times 10^{-19} \, cm^{-1}; \; 1/\mu_\nu = 1.2 \times 10^{18} \, cm = 1.2 \text{ light years}$$

The actual cross section can be slightly larger than that, if other processes are taken into account, but it does not change the fact that neutrinos with a kinetic energy up to about a GeV can pass through about one light-year of dense matter without interacting at all. The very small cross section makes it very difficult to study neutrinos, requiring a large mass of active detectors and a large intensity of neutrinos to record a small number of neutrino interactions. On the other hand, it makes it possible to study neutrino oscillations with neutrinos produced in an accelerator and traveling long distances, of the order of thousands kilometers, without needing any kind of vacuum pipe or tunnel.

The first observation of anti-neutrinos occurred with the reaction:

$$\bar{\nu}_e + p \to n + e^+ \tag{5.47}$$

which is sometimes called "inverse beta decay"; the measured cross section[4] was

$$\sigma(\bar{\nu}_e p \to ne^+) = (11.0 \pm 2.6) \times 10^{-44}\,\mathrm{cm}^2$$

A neutrino flux $\Phi \approx 5 \times 10^{13}\,\mathrm{cm}^{-2}\,\mathrm{s}^{-1}$ obtained from a nuclear reactor was used to detect neutrinos for the first time.

Although neutrinos are able to ionise, for their extremely low cross section they are not considered as ionising radiation for all practical purposes.

5.11 Passage of Radiation from the Point of View of the Material

The energy deposited by radiation in a material has three effects: ionises the material; increases locally its temperature; generates other radiation, which can be absorbed by the material or escape it. An electron from photoelectric effect may ionise other atoms, but also leaves behind a ionised atom, which receives an electron from the environment and will emit X-rays, which can be absorbed photoelectrically by neighbouring atoms and so on. Charged particles leave tracks of ions in the material, but protons and *mesons* and heavy ions at a sufficiently high energy can also displace atoms from a crystal lattice.

The total amount of radiation energy deposited per unit of mass in a material is measured in Gray (Gy) and is also called *absorbed dose*

$$1\,\mathrm{Gy} = 1\,\mathrm{J/kg} = 6.24 \times 10^{12}\,\mathrm{MeV/kg} \tag{5.48}$$

Of course, only the part of material which is interested by the radiation has to be accounted for in the mass value. This is especially important if the radiation is completely stopped, like in the case of charged particles. Large radiation doses can damage solid material, induce or accelerate chemical reactions, accelerate ageing.

5.12 Particle Detectors

Ions separated by radiation can be transformed directly or indirectly into an electric signal, or in some case into a permanent chemical change to detect ionising

---

[4]Frederick Reines and Clyde L. Cowan, Jr., *Measurement of the free antineutrino absorption cross section by protons*, Phys. Rev. 113 (1959), p 273.

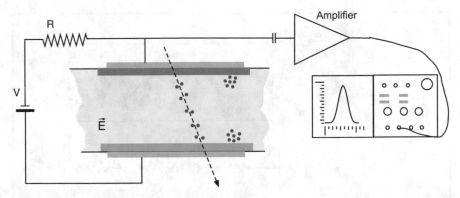

Fig. 5.18 Schematics of a typical solid-state detector. The material is a semiconductor, with a junction on one surface and a highly doped layer on the other surface. The typical thickness is 0.1 to 0.5 mm. The electrodes can also have the shape of strips or pixels, for spatial resolution. The electric charges are separated by the radiation and drift in the electric field of a reverse-biased diode structure. This generates a current, which is amplified by an external circuit and can be observed with an oscilloscope. Positive ions in semiconductor crystals are called *holes*: the lack of electrons move like bubbles in a liquid towards the negatively charged electrode

radiation. This happens in X-ray films, where the radiation ionises crystals of AgBr; the subsequent chemical treatment removes bromine, leaving metallic silver as a dark area in the film only in the areas which have been exposed to radiation.

In solid-state detectors, charged particles lose their energy, which partially goes into crystal excitations, i.e. phonons, and partly to generate electron-hole pairs. On average 3.6 eV is needed for a single pair in silicon, and a signal of $\approx 3 \times 10^4$ electrons can be generated by a minimum ionising particle. Electrons and positive ions ("holes") drift in an electric field in a reverse-biased diode structure and generate a current pulse, which is amplified electronically, as sketched in Fig. 5.18. In this case the distribution of pulse heights is proportional to the distribution of energy loss. For thin detectors it is a modified Landau distribution.

In gas-operated detectors the same mechanisms are in place, but the energy loss in a gas is typically smaller than in solids, because of the density factor. A smaller electric charge is produced. To have a detectable electric signal the method of charge amplification with gas discharge is used: the primary electrons, i.e. those generated by radiation, are drifted to a small sized electrode, like a tiny wire; in its vicinity the electric field is large enough that electrons acquire energy to ionise other molecules of the gas. The resulting electric signal can be furtherly amplified with external electronic circuits. When the internal amplification factor, which is controlled by the voltage and diameter of the wire and by the gas pressure, is relatively small the signal amplitude is proportional to the initial ionisation. The detector is called a *proportional chamber*. When the gas gain is large we have the Geiger counters.

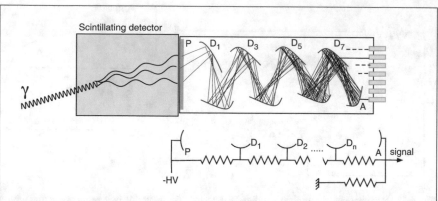

Fig. 5.19 Schematics of a scintillator coupled to a photomultiplier: a scintillator detector (S) produces flashes of light when hit by ionising radiation; light is transmitted through a window to the vacuum tube where a photocathode (P) emits an electron for every photon; electrons are amplified by secondary emission in the various dynodes (D_n); the electric signal is collected at the anode (A)

The Geiger tube is named after Hans Geiger (1982–1945) from Germany, who worked from 1907 to 1912 with E. Rutherford in Manchester, UK and then taught in Tubingen and Berlin, Germany.

A third important class consists of scintillating detectors: all materials emit photons when they are ionised and the electric charges recombine; in the majority of cases the light is also re-absorbed; in scintillator materials this light can be transmitted over a large distance and collected by a light-sensitive electronic device, to be transformed into electric signals. In general, scintillating detectors are made of two-component materials: a light-transparent material which hosts a lower concentration of photoluminescence centres that emit at a wavelength which is not attenuated by the hosting material. An example is thallium in sodium iodide crystal. The energy from primary electrons activates the photoluminescence centres, which produce very small flashes of light, which is detected by a device called a *photomultiplier*, as sketched in Fig. 5.19. Scintillating detectors can be solid inorganic, as NaI(Tl) or solid organic, like plastic scintillators, or liquid. Scintillators are also used to enhance the sensitivity of photographic films to X-rays, as shown in Fig. 5.22.

Detectors can be used to measure particle energies, or the position where particles hit the detectors, e.g. for imaging, or also the time when particles hit the detector. For each application there are advantages and disadvantages of the various technologies. One parameter which is common to all detectors is the efficiency, which is calculated as the fraction of incident radiation which gives an electric signal which is recognised as such. Energy resolution is an important parameter, especially for spectroscopy. Particle detectors are used to measure the energy of the radiation emitted by radionuclides, which is their identifying "fingerprint". In this case the energy resolution is a very important parameter.

5.13 Biological Effects of Radiation

Energy deposited by radiation in living cells has the same effects of creating ions, increasing the temperature locally and generating other radiation. Ionising the saline solution inside cells or the water between cells creates free radicals, which are chemically very reactive. Their presence may have several effects: from killing cells to generating mutations, including carcinogenic ones. The effect of radiation in a living tissue is measured by the absorbed energy weighted by a factor that takes into account how this energy is distributed. This is the *equivalent radiation dose* and is measured in Sievert (Sv), The factor W_R depends on the radiation type and ranges from 1, for gamma rays, to 20 for α particles and heavy ions:

$$\text{Equivalent radiation dose} = \text{absorbed dose} \times W_R; \qquad (5.49)$$

$$1\,\text{Sv} = 1\,\text{Gy} \times W_R$$

Living cells have natural repair mechanisms to the radiation effects, it is also very important to consider the time interval in which this dose is delivered, which is the *equivalent dose rate*, in Sievert per hour (Sv/h) or per year (Sv/y). The same dose has the highest effects if it is delivered in a short time. Radiation is used to kill germs and sterilise equipment and food. It is also used to kill cancer cells: radiotherapy uses alpha and beta-emitting radionuclides linked to molecules which are attracted to cancer cells; hydrotherapy uses collimated beams of charged particles with tunable energy, so that the position of the Bragg peak corresponds to the depth of the tumour. The same radiation, if administered to healthy organs, will kill their cells in exactly the same way and may also cause cancer. An equivalent dose of 1 Sv corresponds to a 5.5% probability of developing cancer from radiation.

We are constantly exposed to natural radiation, from cosmic rays, from radon gas, from other radioactive minerals, like ^{40}K. In general, we absorb an equivalent dose of approximately 1.5–5 mSv/y from natural sources, depending on the location where we live: it is mainly linked to geology and altitude. In Europe the limit on the effective dose for occupational exposure is presently 20 mSv/y. For all manipulation of radioactive sources or operation of X-ray equipment the principle of reducing all unnecessary exposure must be applied.

Other units of exposure are the *rem*, which originally stands for Röntgen equivalent for men: 1 rem = 0.01 Sv. It is now based on the *rad*: 1 rad = 100 Gy. These units are presently officially used only in the USA.

5.14   Radiography

To obtain X-ray images of objects or parts of our body (Fig. 5.20), the object is placed between a X-ray tube and an imaging detector, which can be a special photographic film or an electronic detector. What we see is the shadow of the object projected onto the detector. At a given X-ray energy or wavelength, materials of different Z have different absorption coefficients: while water and other tissue can be transparent, bones which contain calcium absorb the X-rays and project a shadow onto the film.

The ideal generator would produce single-energy X-rays from a point-like source at large distance from the object, in such a way that the rays can be considered as parallel to each other, and objects project a sharp "shadow". The ideal detector has

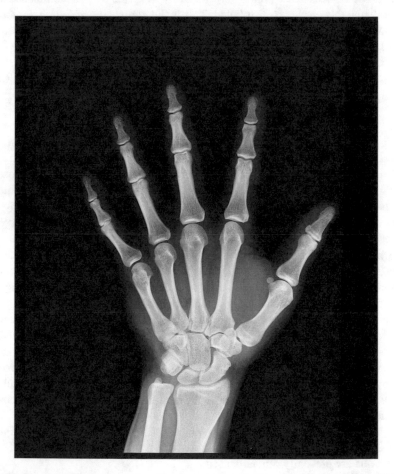

Fig. 5.20 Shadow of a human hand projected on an X-ray detector. The part in black the film has been hit by X-rays, while the parts which have remained white have been shadowed by the bones, which have absorbed the X-rays

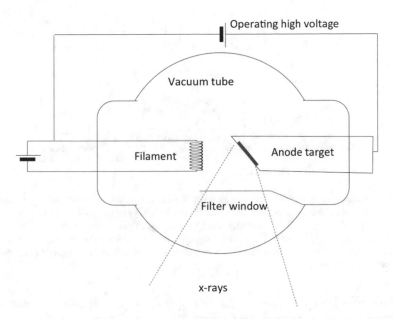

Fig. 5.21 Schematics of an X-ray tube. Electrons are produced by an incandescent filament in vacuum; the production process is called *thermionic emission*; they are accelerated with the high voltage to the desired energy. When they hit the cathode they produce a continuum spectrum due to bremsstrahlung, but they also ionise the cathode atoms. When the electron energy is larger than the K energy levels, occupied by the innermost electrons, also the characteristic X-ray fluorescence lines of the cathode material are emitted. These lines are monoenergetic. Depending on the application, the lower part of the spectrum may be intentionally suppressed by a filter, to reduce the dose to the patient and to minimise the energy dispersion. The cathode is heated by the electron current, which may reach 1 A. A rotating cathode is sometimes used for high intensities. The cathode is inclined, because the X-ray emission is maximum in the direction perpendicular to the electron velocity. To avoid penumbra effects on the film, the electrons are focused to reduce the size of the X-ray source

100% efficiency, a granularity of the order of micrometres and a linear response over several orders of magnitude. The aim is to have an excellent resolution, maximum contrast, while minimising the radiation dose to the patient. In reality an equipment is the best available compromise. Operating voltages range from 25 to 140 kV, anodes are typically made of tungsten, for general purpose, or molybdenum for soft X-ray mammography. Their K-level emission lines are $E_\gamma = 69.5$ and 20 keV, respectively. Currents can reach 1 A and the lower part of the spectrum ($E \leq E_\gamma$) is filtered out to minimise the dose to the patient.

The detector can be a simple, but specific, photographic film, which may have an enhanced detection efficiency by using a scintillator device, which converts X-rays into light. Digital radiography uses either a light-sensitive device to detect scintillation light, or pixelated detectors which convert directly X-rays into electric signals and then into an image. This gives immediate feedback, without any chemical processing of the film. A sketch of the devices is shown in Figs. 5.21 and 5.22.

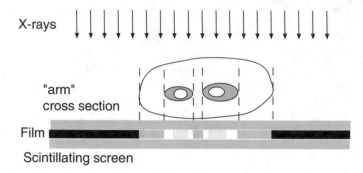

Fig. 5.22 Cross section sketch of an arm in contact with a radiology film detector. Two layers of scintillating material are used to enhance the sensitivity to X-rays of the photo-sensitive film. Black areas correspond to areas which were exposed to the X-rays, light areas are those in the shade of material with high Z, like bones. The contrast depends on the difference of absorption coefficients (μ) of the various materials at the dominant X-ray energy. Electronic detectors can replace the film and lower the dose to the patient

   The unavoidable Compton scattering inside the object, or patient, deflects the incoming X-rays and reduces their energy, adding a noise component to the image and decreasing the contrast. Special collimators can be used to reduce this effect, at the price of efficiency, while energy-sensitive detectors can be used to filter out Compton-scattered photons.

5.15 Problems

5.1 A case study: an X-ray tube operates in the range 25–140 kV. Calculate the range of 40 keV electrons hitting a silver anode. ($\rho = 10.5$ g/cm^3, $M = 107.87$ g/mol). For this study we can in first approximation neglect the energy lost by radiation. Will the obtained value be larger or smaller than the real one? Assume a current of 10 mA, calculate the energy loss in 1 s in the anode and compare this value with the purely electrical power. Assuming no heat exchange, a mass of the anode $m_a = 20$ g, and a heat capacity $C_h = 25$ J/mol/K, calculate the temperature of the anode after 10 s of continuous operation.

5.2 A radiology lab uses 40 keV X-rays from an apparatus with a silver anode, whose measured emission spectrum is shown in Fig. 5.23. The walls of the lab need to be thick enough to contain the radiation. What thickness of lead is needed to attenuate the radiation intensity to 0.1% of the initial value?
   What thickness is needed if we use concrete instead?.
The energy of the Ag emission lines K_β and K_α and the corresponding attenuation coefficients for lead are reported in Table 5.1. The lead density is 11.3 g/cm^3. Should we use the k_β or the spectrum end-point energy? The attenuation coefficient for concrete at 40 keV is 0.5 cm^2/g and its density is 2.4 g/cm^3.

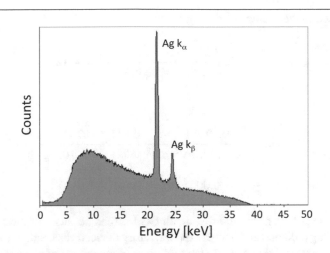

Fig. 5.23 Spectrum of X-rays emitted by a tube with silver anode, operated at 40 kV and measured with a CdTe detector (Courtesy Amptek, Inc. www.Amptek.com)

Table 5.1 Table of the X-ray attenuation coefficients in lead, for three values of the X-ray energy

Line	E (keV)	$\mu(\mathrm{cm}^2/\mathrm{g})$
k_α	22	70
k_β	25	45
Bremsst.	40	14

5.4 Calculate the range of 5 MeV α particles ($z = 2$) in nitrogen at standard conditions, which is an approximation for air, with density $\rho = 1.2\,\mathrm{mg/cm}^3$, $Z = 7$ and atomic mass $M = 14.00$ g/mol. The mass of an α particle is $m_\alpha = 3727.3\,\mathrm{MeV/c}^2$.

5.5 A radioactive source of ^{90}Sr emits β^- with an end-point of 546 keV. Calculate the range of electrons in aluminium, neglecting radiative energy losses. What thickness of aluminium would shield 99% of all radiation, including bremsstrahlung? Aluminium: $M = 26.98$ g/mol, $\rho = 2.7\,\mathrm{g/cm}^3$.

5.16 Solutions

Solution to 5.1   First of all we verify that the kinematics of 40 keV electrons can be approximated by non-relativistic formulae, so Eq. (5.41) is valid. From Eq. (5.41) we first calculate the average ionisation energy $I = 22.8 + 47 \times 9.7 = 480$ eV to be used in the average of the log term:

$$L = \ln \left(\frac{2 \times 511 \times 40 \text{ keV}^2}{511 \times 0.48 \text{ keV}^2} \right) = 4.42$$

we have:

$$R = 0.5 \times \frac{0.0016 \text{ MeV}^2}{0.078 \text{ MeV}^2 \text{ cm}^2} \frac{0.511}{0.511} \frac{107.00 \text{ g/mol}}{47 \times 10.5 \text{ g/cm}^3}$$

$$\times \frac{1}{4.42} = 0.0005 \text{ cm} = 5 \,\mu\text{m}$$

This value is larger than the real depth of electrons because we have neglected the braking radiation and because the trajectory of electrons inside the material is not a straight line. However, we have an order of magnitude of a few micrometres of the anode which takes part to the X-ray emission. 10 mA= 10^{-2} C/s= $10^{-2}/1.6 \times 10^{-19} = 6.25 \times 10^{16}$ e/s. In 1 s the energy loss is

$$6.25 \times 10^{16} \times 40 \times 10^3 = 250 \times 10^{19} \text{ eV} = 250 \times 10^{19} \times 1.6 \times 10^{-19} = 400 \text{ J}$$

The value of the electrical energy is

$$W = I V t = 10^{-2} \text{ A} \times 40 \times 10^3 \text{ V} \times 1 \text{ s} = 400 \text{ J}$$

The temperature of the anode increases by ΔT

$$\Delta T = t \times \frac{W m_a}{M C_h} = 10 \text{ s} \times \frac{400 \times 10}{107.87 \times 25} = 14 \text{ K}$$

Even with a very modest current, the temperature of the anode increases considerably. This is the reason why rotating or water-cooled cathodes are used in industrial applications.

Solution to 5.2   As the photoelectric cross section for a given material decreases with the cube of the X-ray energy, we consider the maximum energy in the spectrum, which is 40 keV. The attenuation coefficient for lead is $\mu' = \rho\mu = 11.3 \times 14 = 158.2 \text{ cm}^{-1}$.

$$N(x)/N(0) = e^{-\mu' x} = 0.001; \quad \Rightarrow -\mu' x = \ln(0.001);$$

$$x = -\frac{1}{\mu'} \ln(0.001) = 6.91/158.2 = 0.4 \text{ mm}$$

In case of concrete $\mu' = \rho\mu = 2.4 \times 0.5 = 1.2 \text{ cm}^{-1}$ and

$$x = -\frac{1}{\mu'} \ln(0.001) = 6.91/1.2 = 5.7 \text{ cm}$$

Fig. 5.24 Energy loss profile of 5 MeV α particles in air, as calculated with a numerical integration of Eq. (5.39)

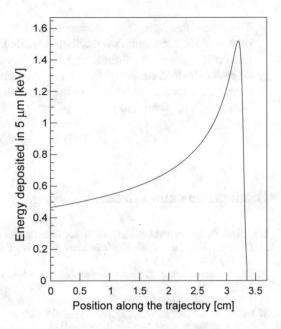

Solution to 5.4   From Eq. (5.41) we first calculate the average ionisation energy $I = 22.8 + 7 \times 9.7 = 90.7$ eV to be used in the average of the log term:

$$L = \ln\left(\frac{2 \times 0.511 \times 5 \text{ MeV}^2}{3727.3 \times 9 \times 10^{-5} \text{ MeV}^2}\right) = 2.86$$

we have:

$$R = 0.5 \times \frac{25 \text{ MeV}^2}{0.078 \text{ MeV}^2 \text{ cm}^2} \frac{0.511}{3727.3} \frac{14.00 \text{ g/mol}}{7 \times 1.2 \times 10^{-3} \text{ g/cm}^3} \times \frac{0.25}{2.86} = 3.2 \text{ cm}$$

Alpha particles are absorbed by a few centimetres of air (Fig. 5.24).

Solution to 5.5   From Eq. (5.41) we first calculate the average ionisation energy for aluminium: $I = 22.8 + 13 \times 9.7 = 149$ eV to be used in the average of the log term:

$$L = \ln\left(\frac{2 \times 0.511 \times 0.546 \text{ MeV}^2}{0.511 \times 1.49 \times 10^{-4} \text{ MeV}^2}\right) = 8.89$$

we have:

$$R = 0.5 \times \frac{0.298 \text{ MeV}^2}{0.078 \text{ MeV}^2 \text{ cm}^2} \frac{0.511}{0.511} \frac{26.98 \text{ g/mol}}{13 \times 2.7 \text{ g/cm}^3} \times \frac{1}{8.89} = 0.33 \text{ cm}$$

The β particles from ^{90}Sr are stopped by a few mm of aluminium. In addition, they produce bremsstrahlung radiation, with end-point at 546 keV and characteristic K-lines of aluminium, at 1.6 keV. The low-energy X-rays from Al can be easily shielded either by aluminium itself or by a thin layer of shielding with higher Z. To shield at 99% from bremsstrahlung we can consider to use iron, with $\rho = 7.86$ g/cm^3 and $\mu = 0.084$ cm^2/g.

$$-\mu' x = \ln(0.01) \Rightarrow x = 4.6/0.66 \approx 7 \text{ cm}$$

Bibliography and Further Reading

L. Cerrito, *Radiation and Detectors* (Springer International Publishing, Singapore, 2017)

A. Del Guerra, *Ionizing Radiation Detectors for Medical Imaging* (World Scientific, Singapore, 2004)

R. Fernow, *Introduction to Experimental Particle Physics* (Cambridge University Press, Cambridge, 1986)

G.F. Knoll, *Radiation Detection and Measurement* (Wiley, New York, 2012)

W.R. Leo, *Techniques for Nuclear and Particle Physics Experiments* (Springer, Berlin, 1993)

B.R. Martin, *Nuclear and Particle Physics - An Introduction*, 2nd edn. (Wiley, Chichester, 2009)

B. Povh et al., *Particles and Nuclei: An Introduction to the Physical Concepts* (Springer, Berlin, 2015)

E. Segré *Nuclear and Particle Physics* (W. A. Benjamin, Reading, 1977)

M.G. Stabin, *Radiation Protection and Dosimetry* (Springer Science+Business Media, New York, 2007)

S. Tavernier, *Experimental Techniques in Nuclear and Particle Physics* (Springer, Berlin, 2010)

J.E. Turner, *Atoms, Radiation, and Radiation Protection* (Wiley WCH, Weinheim, 2007)

Chapter 6
Introduction to Particle Physics

6.1 Introduction to Fundamental Interactions

We have experience from everyday life of the electromagnetic and the gravitational interactions. Their form is similar, but the relative strength is really different.

$$F = G_N \frac{m_1 m_2}{r^2}; \qquad\qquad F = \frac{1}{4\pi\epsilon_0}\frac{q_1 q_2}{r^2}; \qquad (6.1)$$

They are both long-range forces, they have a nice "geometrical" variation with $1/r^2$, which is very convenient in a 3-dimensional world. The mass plays the role of "gravitational charge". This is the equivalence principle of *general* relativity. The ratio of the two forces for an electron and a proton, at any arbitrary distance, which cancels out in the ratio, is 10^{-39}. So gravity is completely negligible in particle physics. There is no theory of quantum gravity yet. Let's now look at other forces. From the structure of the atomic nucleus, it is clear that there must exist a force which binds together protons and neutrons. This force has to be stronger than the electrostatic repulsive force between protons, which are all positively charged. It is called *"strong nuclear force"*. Its effects are negligible in ordinary matter. This means that the strong nuclear force must have a limited range, within the size of the nucleus, which is a few femtometres in radius (1 fm = 10^{-15} m) (Fig. 6.1).

© Springer Nature Switzerland AG 2018
S. D'Auria, *Introduction to Nuclear and Particle Physics*,
Undergraduate Lecture Notes in Physics,
https://doi.org/10.1007/978-3-319-93855-4_6

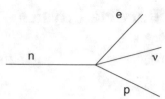

Fig. 6.1 Initial (1934) model of the weak interaction involving four particles. The β decay of the neutron is shown in the $t - x$ plane. Time is along the horizontal axis. This type of interaction is not mediated by any force. All particles are in contact at the interaction time, and it is called "4-fermion contact interaction". We now know that the β decay is mediated by the W boson

Nuclides decaying β emit particles in the 0.1–1 MeV energy range with a continuous distribution, thus indicating the emission of a second particle, which escapes detection; this particle does not ionise matter and, therefore, has to be electrically neutral. The other neutral particle we have met so far in this book is the *neutron*, which was discovered by James Chadwick in 1932. Neutrons decay with an average lifetime of about 14′40″. The decay products are a proton, an electron and "something else", which must be also neutral, and have a small mass. This particle was called *neutrino*, which stands for "the little neutral" particle. It escaped direct detection until 1956. The fact that this particle escapes detection means that its interaction with other particles, in particular electrons and protons, has to be particularly weak. Hence the name of *weak nuclear force*. The relatively long lifetime of the neutron is also an indication of the weak force involved in its decay. The neutrino is our probe to the weak force, which is so weak that neutrinos can traverse the planet earth without interacting. The weak force does not produce any bound state of the matter. It is only visible in decays and is universal: all the spin ½ fundamental particles that we know of are subject to the weak force.

The (anti) neutrino was experimentally discovered in 1956 by Clyde Cowan, Frederick Reines et al. using as a source the Savannah River nuclear plant, in South Carolina, USA. Reines was awarded the Nobel prize in 1995.

The *strong nuclear* force keeps together quarks, to form particles which we call *mesons* and *baryons*. Protons and neutrons are the only baryons we have met so far. Neutrons are our probe to the nuclear force: they are neutral and do not interact, or interact very weakly, with the long-range electromagnetic force. Its *strong* interaction is the residual interaction of a more fundamental one, the quantum chromo dynamics (QCD), which holds together the components of neutrons and protons: the *quarks*. The fundamental forces are listed in Table 6.1.

Table 6.1 Comparison of the apparent strength of the fundamental forces, relative to the electromagnetic force

Force	Strength	Carrier	Symbol	Mass
Electromagnetic	1	Photon	γ	0
Weak	10^{-5}	Vector bosons	$Z^0, W^{\pm}$	91–80
Strong	10^3	Gluons (8)	g	0
Gravitation	10^{-39}	Graviton?	G	0

The mass of the force carriers is in GeV/c². There is no experimentally established quantum theory that includes gravity, so the existence of graviton is still hypothetical

> It should be noted, in passing, that neutrons, even though are electrically neutral, have a "residual" electromagnetic interaction: its magnetic moment is $\mu_n = -1.913 \times \mu_N = -1.913 \times e\hbar/2m_p$. This is explained by the fact that it is a composite particle, made by quarks, which are electrically charged.

Forces are mediated by fields. The reader may be familiar with the electric and magnetic fields. In every point of our laboratory space we can define a quantity, actually a "little arrow", a vector, indicating the electric field $\vec{E}$ in that point, and the same for the magnetic field $\vec{B}$. The charged particles feel a force, which is proportional to the value of the field in that point. We also know that rapidly varying electromagnetic fields are radio waves and light. In this example the electromagnetic force is mediated by the corresponding fields $(\vec{E}, \vec{B})$, which can be represented in relativistic notation by a single field $\mathbf{A(x)}$, the 4-vector potential. There is a quantum particle corresponding to this field and it is the *photon*. It has zero mass, and therefore in vacuum it always travels at a constant speed, the speed of light. Similarly, the strong and weak interactions have a corresponding field, which results into particles which carry, or mediate, the force. The weak interaction is mediated by two particles, which are massive : the $W^{\pm}$ and the Z^0. The strong interaction is mediated by the *gluons*, which are massless, like the photon, but are "prisoners of their strength": they cannot exist as free particles, they are *confined* inside other, composite, particles.

6.2   Strength and Range of the Interactions

We'll now make some quantitative statements about the strength of the fundamental forces, trying to avoid, if possible, to purely put them into categories. In doing so we have to consider not only the strength but also the range of a force. The electromagnetic force has infinite range. We'll see that the *strong* and the *weak* nuclear force have a limited range, but for very different reasons. We'll need some results from quantum mechanics to be more quantitative.

The Heisenberg uncertainty relations link the energy available for a decay to the lifetime of the corresponding particle. From the general Heisenberg uncertainty relation:

$$\Delta E\, \Delta t \geq \frac{\hbar}{2} \tag{6.2}$$

we can define the *decay width* Γ as

$$\Gamma = \frac{\hbar}{2\tau} \tag{6.3}$$

where τ is the average lifetime of the decay. A stable particle has a very definite value of its mass. The mass of unstable particles can only be measured in a finite time interval of their existence before they decay. As time and energy are conjugate variables, they are subject to the quantum mechanics uncertainty. The mass of a particle is its energy when it is at rest, so also the mass of unstable particles has an intrinsic range of variation. We measure the mass of unstable particles by measuring the energy and momenta of the decay products and calculating the invariant mass, as in Eq. (2.39). When repeating the experiment several times, the resulting values of the invariant mass will not be all the same. One component of uncertainty is the experimental error, but even when this is reduced to a minimum value, the distribution of masses is not a narrow line, but a bell-shaped curve, which is not a Gaussian, and whose width is larger when the lifetime is shorter. The width of this curve is the intrinsic variation of the mass and is indicated with Γ. Its dimensions are the same as an energy.

In particle physics it is usual to measure masses in MeV/c^2 or GeV/c^2. As an example, the Z^0 has an average value of the mass of (91.188 ± 0.002) GeV/c^2 and a width of (2.495 ± 0.002) GeV/c^2, or 2.7% of the mass value, as shown in Fig. 6.2.

Considering unstable particles, we would expect that particles of approximatively the same mass would also have about the same average lifetime, if the decay occurs with only one fundamental interaction, because the energy available to the final states is about the same. In Fig. 6.3 we can see that the lifetimes of these mesons vary by 15 orders of magnitude. Three interactions play their role, with different strength, and, therefore, with different decay times, which are longer for weaker interactions. The weak interactions are characterised experimentally by relatively large lifetimes,[1] which are of the order of 10^{+3} (neutron) to 10^{-12} s. By comparison, the electromagnetic decays have lifetimes of 10^{-18} to 10^{-20} s and the strong decays have lifetimes of the order of 10^{-22} s. Of course, several corrections must be applied to this "rule of thumb", but this is an indication of the various

---

[1]The most notable exception to the large lifetimes rule is the *top* quark, which decays weakly, but very rapidly, due to its very large mass.

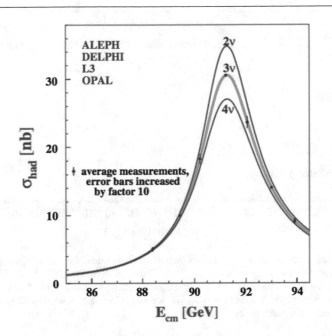

Fig. 6.2 The intrinsic width of the Z^0 boson has been measured very precisely by measuring its production cross section in $e^+ e^-$ collisions at various energies. The width of the Z^0 depends on the total number of light neutrino species existing, so this figure which illustrates the natural width of a particle is also the experimental proof that there are just three types of light neutrinos (from Phys. Reports, 427, 5–6 (2006), p. 277)

"strength" of the forces that are involved, or one of the motivation to introduce three different forces. In general, reactions can be divided into scattering

$$A + B \rightarrow C + D; \quad 2 \text{ initial } \rightarrow 2 \text{ final}, \tag{6.4}$$

where A, B, C and D are particles, and decays, like

$$B \rightarrow C + D \tag{6.5}$$

In both cases we want to calculate the transition rate: in the first case we calculate the *cross section*, σ in the other the lifetime, or the *decay width* $\Gamma = \hbar/\tau$.

There is a general rule to calculate the rate for these reactions to occur, factorising the components. The reaction rate in case of a $2 \rightarrow 2$ reaction depends on the flux

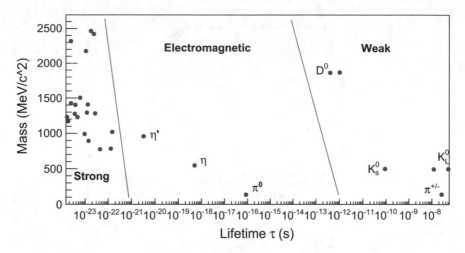

Fig. 6.3 Lifetime of some *mesons* as a function of their mass. Lifetimes span 15 orders of magnitudes. Strong decays occur with very short lifetime in the range 10^{-21} to 10^{-23} s, and only have lighter mesons as decay products. Electromagnetic decays have lifetimes of the order of 10^{-18} to 10^{-21} s and have photons in the final state. Weak decays have, in general, a much longer lifetime, and include leptons in the possible final states

J_A of the incident particles and on the number density N_B of the target particles, as we have seen before:

$$W_r = J N \sigma_r = I \frac{N_A}{M_A} \rho \, \delta x \, \sigma_r \qquad (6.6)$$

where ρ is the density of the target, N_A the Avogadro's number, M_A the atomic weight in a.m.u. , or Dalton, Eq. (7.10), and δx the thickness of the target. The quantity JN is called the *Luminosity* and it gives the rate, once it is multiplied by the effective cross section σ_r.

In the case of a decay (Eq. (6.5)), the decay rate, as shown in Chap. 4, depends on the initial number of particles N and on the mean lifetime $\tau = \hbar / \Gamma$. We want to calculate the decay width and/or the scattering cross section. Grouping the two cases, we calculate the transition rate from an *initial* to a *final state*. This calculation can be factorised: one part depends only on the initial and final states and one part is specific to the interaction. The mathematical space of coordinate and momenta of a given physical system is called the *phase space*. For a free, point-like particle it is a six-dimensional space $\mathbb{R}^6$. The term linked to the final states is purely quantum origin: the transition rate depends on the *density of the phase space* which is available to the decay products. It depends on the energy available to the final states. Let's call this factor ρ_f. When considering a decay, if there is not enough energy available to the final products, the phase space term is zero. For a decay we can write

$$\Gamma = \rho_f |\mathcal{M}_{if}|^2 \qquad (6.7)$$

The term which depends on the nature of the interaction is called "*the square of the transition amplitude*" $|\mathcal{M}_{if}|^2$ from the *initial (i)* to the *final (f)* state

$$W_r = JN \frac{2\pi}{\hbar} \rho_f |\mathcal{M}_{if}|^2 \qquad (6.8)$$

or

$$\sigma = \rho_f |\mathcal{M}_{if}|^2 \qquad (6.9)$$

This term takes into account the interaction strength. The equations above (Eqs. (6.9) and (6.7)) are called the *Fermi's golden rule*. We'll see a qualitative application of it when considering the $\beta^\pm$ decays. The argument of linking the decay time with interaction strength can now be made more quantitative: the phase space factor depends on the mass of the particle, which is about constant, so the variation is given by the amplitude, which is different for different interactions. We have already encountered a cross section calculation, the K-N cross section for the Compton scattering (Eq. (5.34)), which is an electromagnetic process. The amplitude term contains the electric charge, or the fine structure constant, and the angular variation.

Fermi's golden rule, although was originally found by P.A.M. Dirac, is named after Enrico Fermi, 1901–1954, Nobel prize in 1938, who stressed the importance of this formula.

For a generic interaction, we assume that it is invariant for time reversal (T) and parity (space inversion) (P) we have that

$$|\mathcal{M}_{if}|^2 = |\mathcal{M}_{fi}|^2. \qquad (6.10)$$

This is called the *principle of detailed balance*. It was experimentally confirmed for the strong interaction, as shown in Fig. 6.4. The weak interaction violates T-invariance, but to a first approximation detailed balance still holds true for the weak interaction. At this point we want to calculate $\mathcal{M}_{if}$. The calculation is easier if we are allowed to consider the interaction as a small perturbation to the propagation of free particles. We cannot do so in some cases, for instance, for the strong nuclear force at low energies. However, this method works very well for the electromagnetic and weak interactions, giving results that match the experimental values. To quantify the concept of "small perturbation" we can introduce the so-called structure constants. We have already introduced the *fine structure constant*

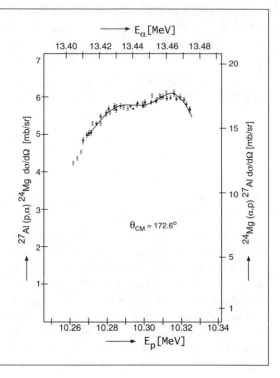

Fig. 6.4 Experimental confirmation of the detailed balance: the reaction $^{24}Mg + \alpha \rightleftharpoons {}^{27}Al + p$ can proceed both ways. After normalising for the experimental set-up and for the phase space factors, the differential cross section $d\sigma/d\Omega$ of the two reactions has the same dependence on the energy of the incoming particle (p or α). Reprinted with permission from W. von Witsch, A. Richter, and P. von Brentano, Phys. Rev. Letters 19, n. 9 (1967) copyright (1967) by the American Physical Society

for the electromagnetic interaction (Eq. (5.37)):

$$\alpha_{em} \equiv \frac{e^2}{4\pi\epsilon_0\hbar c} \approx \frac{1}{137.04} = 7.297 \times 10^{-3} \qquad (6.11)$$

This number is also called a *coupling* and, so long as it is well below the numerical value of 1, we can use *perturbation theory*. In practice, it is like approximating a function with the first or second term of its Taylor series. The terms of the series are proportional to powers of the coupling of the interaction under consideration.

The purpose of the following arguments is to show that the weak nuclear force has the same strength as electromagnetic force, although, at a first sight, they seem so different.

First of all, we have to guess a potential for the new short-ranged nuclear interactions. The nuclear force has been shown to have a range beneath which its effects are negligible. Hideki Yukawa (Fig. 6.5), back in 1935, introduced a potential which modifies the $1/r$ shape, which is typical of a Coulomb potential, with an exponential function. The new term reduces the effects of the corresponding force at a distance larger than a "characteristic distance" which we call R.

$$V(r) = \frac{g^2}{4\pi}\frac{1}{r}e^{-r/R} \qquad (6.12)$$

Fig. 6.5 Hideki Yukawa, 1907–1981, Nobel prize in 1949 for having predicted the existence of the π meson. He was born and educated in Japan. His name was Ogawa, he adopted his wife's family name, as she had no brother. Photo from April 1951, Asahi Shimbun newspaper

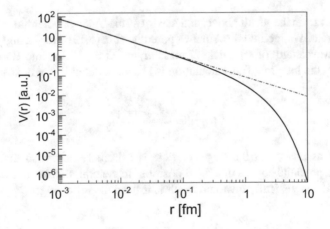

Fig. 6.6 The Yukawa potential (solid line) is qualitatively compared to the Coulomb potential (dashed line). The two have the same shape for $r < R \approx 1$ fm, but the Yukawa potential goes exponentially to zero at large r. Note the logarithmic scale on both axes

where g is the "charge" of the interaction and R is the *range* of the force. A qualitative comparison between the Yukawa and the Coulomb potentials is shown in Fig. 6.6. In his original article Yukawa applied this principle to the strong interaction, which experimentally has a range $R \approx 10^{-15}$ m. In physical terms, the quantum of electromagnetic interaction is the photon, which is massless. Suppose that the quantum that mediates an interaction has a mass m, then it can only appear for a time Δt which is limited by the uncertainty principle (Eq. (6.2)):

$\Delta t \leq \hbar/(2mc^2)$. In other words, the interaction has a range R

$$R \approx c\,\Delta t = \frac{\hbar c}{2m\,c^2} \tag{6.13}$$

This is the "Compton wavelength" of a particle of mass m. If we use the experimental value of the radius of a proton $R \approx 10^{-15}$ m, then the mass of the mediator of the force is $mc^2 = \hbar c/(2R) \approx 100$ MeV. There is indeed a particle, which is called a *pion* and is denoted by the symbol $\pi^{\pm,0}$, whose mass is 140 GeV$/c^2$. Thus the pion was assumed to be the fundamental particle mediating the strong force. The pion can be used to describe the $p-n$ scattering, but we now know that it is not a fundamental particle. The strong interaction is more complicated than that. These ideas, or heuristic arguments, however, turned out to be applicable to the weak interactions, as follows. In quantum perturbation theory the Born[2] approximation relates the amplitude of the probability for a particle to scatter off a potential $V(r)$ to the Fourier transform of the potential:

$$\mathcal{M}(\vec{q}) = \int d^3 r\, V(\vec{r}) e^{i\frac{\vec{q}\vec{r}}{\hbar}} \tag{6.14}$$

We can start from the result above, and use for V the Yukawa potential to obtain the transition rate. We include the Yukawa potential (Eq. (6.12)), replacing R with the Compton wavelength of a particle of mass m_x, as in Eq. (6.13) into Eq. (6.14) and we perform the integral. This calculation is shown in detail in the next section. We obtain

$$\mathcal{M}(q^2) = \frac{g^2 \hbar^2}{|\vec{q}|^2 + m_x^2 c^2} \tag{6.15}$$

This result assumes that the target particle is extremely heavy, so that it can be treated as a potential source in a fixed position. In general, this is not the case. The correct and relativistically invariant result is instead the following:

$$\mathcal{M}(\mathbf{q}^2) = \frac{g^2 \hbar^2}{|\mathbf{q}|^2 - m_x^2 c^2} \tag{6.16}$$

The denominator of the equation above (Eq. (6.16)) is called the *propagator*. The low-energy limit ($q \to 0$) of the above formula is a constant, which we call G_F (Fermi's constant):

$$-G_F = \frac{g^2 \hbar^2}{m_x^2 c^2} \tag{6.17}$$

---

[2] After Max Born, 1882–1970, from Germany, he taught in Göttingen, Cambridge and Edinburgh and was awarded the Nobel prize in 1954, for his fundamental research in quantum mechanics, especially in the statistical interpretation of the wave function.

In the right-hand side of the equation above there are two quantities which we do not know: the "coupling" g and the mass of the mediating particle m_x. However, the value of G_F can be calculated from the experimental value of the muon lifetime. We actually calculate $G_F/(\hbar c)^3$, to obtain the correct units [Energy]$^{-2}$. We define a "structure constant" for the weak interaction and we call it α_W, along the lines of the fine structure constant α_{EM} (Eq. (6.11)):

$$\alpha_W = \frac{g^2}{4\pi\hbar c} \tag{6.18}$$

This is a dimension-less constant, which we can use in the definition of G_F instead of g

$$G = \frac{G_F}{(\hbar c)^3} = \frac{\hbar^2}{\hbar^3 c^3} \frac{g^2}{m_x^2 c^2} = \frac{4\pi}{m_x^2 c^4} \frac{g^2}{4\pi\hbar c} = \frac{4\pi\alpha_W}{m_x^2 c^4} \tag{6.19}$$

From the experimental lifetime of the muon decay we get value of the Fermi's constant:

$$G = G_F/(\hbar c)^3 = 1.166 \times 10^{-5} \text{ GeV}^{-2} \tag{6.20}$$

Now, let's assume, as a heuristic hypothesis that

$$\alpha_{weak} \approx \alpha_{em} = \alpha \approx 1/137 = 7 \times 10^{-3}$$

This means that we can calculate approximately the mass of the mediating particle:

$$m_x c^2 = \sqrt{\frac{4\pi\alpha}{G}} = \sqrt{\frac{4\pi \times 7 \times 10^{-3}}{1.166 \times 10^{-5}}} = 88 \text{ GeV}$$

The mediating particles of the weak force have been experimentally discovered in 1984 and are called $W^{\pm}$ and Z^0 *vector* bosons. They indeed have a mass close to the value which we have calculated "on the back of the envelope". This means that our initial guess $\alpha_W \approx \alpha_{EM}$ was correct: the weak nuclear force has about the same coupling as the electromagnetism. This is why we now speak of *electroweak interactions*, because they are essentially unified.

In summary: there are two ways for an interaction to be weak: to have small values for the coupling constant and/or to be mediated by massive particles, which means it has a short range. The latter is the case of the weak interaction: it is not so weak, per se: it has similar couplings as the electromagnetic interaction, but it is mediated by two massive particles.

The next section is dedicated to show how Eq. (6.15) is calculated from Eqs. (6.12) and (6.14).

6.3 The Yukawa Potential in Born Approximation

In this section we explain how from the Yukawa potential and the Born approximation (Eqs. (6.12) and (6.14)) we obtain the propagator as in Eq. (6.15).

From quantum mechanics: the transition amplitude from a potential is given by

$$\mathcal{M}_{if} = <\psi_{in}|V|\psi_{out}> = \frac{2\pi}{\hbar} \int d^3r \psi_{in}^*(r) V(r) \psi_{out}(r) \qquad (6.21)$$

We consider the case of an incoming plane wave and we assume that the outgoing
wave function is also a plane wave:

$$\psi_{out} = \frac{1}{\sqrt{\text{Vol}}} e^{i\frac{\vec{q}_i \vec{r}}{\hbar}} \ ; \quad \psi_{in} = \frac{1}{\sqrt{\text{Vol}}} e^{i\frac{\vec{q}_f \vec{r}}{\hbar}} \ ; \qquad (6.22)$$

"Vol" indicates the volume in which the wave function is normalised. The case under
study is a Yukawa potential of the type

$$V(r) = \frac{-g^2}{4\pi} \frac{1}{r} e^{-\frac{r}{R}} \qquad (6.23)$$

where R is some characteristic distance, which is given by the uncertainty principle
for a particle of mass M_x:

$$R = \frac{\hbar}{M_x c} \qquad (6.24)$$

Substituting into (6.21) we have

$$\mathcal{M}_{if} = -\frac{1}{\text{Vol}} \frac{g^2}{4\pi} \int d^3r \frac{1}{r} e^{-\frac{r}{R}} e^{-i/\hbar \vec{q} \cdot \vec{r}} \ , \qquad (6.25)$$

where $\vec{q} = \vec{q}_i - \vec{q}_f$. We can use polar coordinates:

$$d^3r = r^2 \sin\theta \, d\theta \, d\phi \, dr$$

and $\vec{q} \cdot \vec{r} = |q|r \cos\theta$ The integral becomes

$$\int \int \int \frac{1}{r} e^{-\frac{r}{R}} e^{-i/\hbar qr \cos\theta} r^2 \sin\theta \, d\theta \, d\phi \, dr \qquad (6.26)$$

$$\int_0^{2\pi} d\phi \int_0^\pi \int_0^\infty r e^{-\frac{r}{R}} e^{-(i/\hbar) qr \cos\theta} \sin\theta \, d\theta \, dr \ ; \text{ but } \sin\theta \, d\theta = d(\cos\theta) \qquad (6.27)$$

$$\int_0^{2\pi} d\phi \int_{-1}^{1} \int_0^{\infty} r e^{-\frac{r}{R}} e^{-(i/\hbar) qr \cos\theta} d(\cos\theta) dr \quad |\text{call} \cos\theta = \mu \qquad (6.28)$$

$$2\pi \int_0^{\infty} dr \int_{-1}^{1} r e^{-\frac{r}{R}} e^{-(i/\hbar) qr\mu} d\mu \qquad (6.29)$$

Now the inner integral:

$$\int_{-1}^{1} e^{-(i/\hbar) qr\mu} d\mu = \qquad (6.30)$$

$$\frac{i\hbar}{qr} e^{-(i/\hbar) qr\mu} \Big|_{-1}^{1} = \frac{i\hbar}{qr} \left(e^{-(i/\hbar) qr} - e^{+(i/\hbar) qr} \right) = \qquad (6.31)$$

$$= \frac{2 \sin (qr/\hbar)}{qr/\hbar} \qquad (6.32)$$

Substituting into (6.29):

$$2\pi \int_0^{\infty} r e^{-\frac{r}{R}} \frac{2 \sin \left(\frac{qr}{\hbar}\right)}{\frac{qr}{\hbar}} dr = \qquad (6.33)$$

$$\frac{4\pi}{q/\hbar} \int_0^{\infty} e^{-\frac{r}{R}} \sin (qr/\hbar) dr = \qquad (6.34)$$

This can be solved by integrating twice by parts, resulting in:

$$\int_0^{\infty} e^{ax} \sin(bx) dx = \frac{e^{ax}}{a^2 + b^2} (a \sin(bx) - a \cos(bx)) \Big|_0^{\infty}$$

where $a = -1/R$ and $b = q/\hbar$; the integral becomes:

$$\frac{4\pi}{q} \left(\frac{e^{-\infty}}{1/R^2 + q^2} (\text{someth. oscill.}) - \frac{1}{1/R^2 + q^2} (-q) \right) = \qquad (6.35)$$

$$- 4\pi \frac{1}{\frac{1}{R^2} + q^2} \qquad (6.36)$$

Using the definition of R (Eq. (6.24)) and substituting Eq. (6.36) into Eq. (6.25), we finally obtain

$$M_{fi} = \frac{g^2 \hbar^2}{q^2 + M_x^2 c^2} \qquad (6.37)$$

which is Eq. (6.15). The volume at the denominator of Eq. (6.25) cancels in the calculation of the cross section with a volume at the numerator, which comes from the phase space factor.

Before looking at the strong force we have a look at the particles, which we believe, at present, are fundamental.

6.4   Fundamental Particles

In this section we'll introduce the twenty-four fundamental particles, plus the corresponding 24 anti-particles, and the twelve fundamental fields which carry the interaction between particles. The particles and fields are primarily classified according their spin. This quantum number tells us how the quantum wave function which corresponds to the particle transforms with space rotations.

6.5   Bosons and Fermions

The intrinsic angular momentum, or *spin*, deeply influences the behaviour of systems made of many indistinguishable particles. The particles that we believe are elementary are point-like; they don't show any spatial extension, as nuclei do. It is convenient to represent them by tiny little arrows, because all these particles carry an intrinsic angular momentum, in half-integer or integer units of $\hbar$. This tells us how the wave function behaves under rotations of the reference frame. The only fundamental particle that carries no angular momentum, that we know so far, is the Higgs boson.

The *spin-statistics theorem* states that all particles which have semi-integer angular momentum follow Fermi-Dirac statistics and it is subject to the Pauli's exclusion principle: no two identical particles can be in the same location of the phase space, or can have the same quantum numbers. These particles are collectively named *fermions*. Fermi-Dirac statistics means, among other things, that the collective wave function of a system of N indistinguishable particles of the same type, e.g. N electrons, changes sign if two particles are swapped:

$$\psi(f_1, f_2, f_3, \ldots, f_N) = -\psi(f_1, f_3, f_2, \ldots, f_N) \qquad (6.38)$$

In the above formula f_i indicates the quantum state of the ith fermion. A consequence of this is that two identical fermions cannot be in the same quantum state: as particles are indistinguishable, if they are in exactly the same state the wave function would be symmetrical, or zero. This is Pauli's exclusion principle. The wave function, or quantum state, of the system is said to be *antisymmetric* under interchange of any two particles. Electrons, protons and neutrons are all fermions.

Fig. 6.7 Satyendranath Bose (1894–1974) from India, where he was born and educated. His article "Planck's Law and the Hypothesis of Light Quanta" was translated to German by Einstein and published in Zeitschrift für Physik 26:178–181 (1924). Bose had previously translated Einstein's article on general relativity into English. He taught at the Universities of Dhaka and Calcutta. Photo taken in Paris, 1925, courtesy AIP Emilio Segrè Visual Archives

All particles that carry integer spin, in units of $\hbar$, or no angular momentum at all, follow Bose–Einstein (Fig. 6.7) statistics: the wave function describing a system of N indistinguishable particles with integer spin does not change sign when swapping two particle indices:

$$\phi(b_1, b_2, b_3, \ldots, b_N) = \phi(b_1, b_3, b_2, \ldots, n_N) \tag{6.39}$$

These particles are not subject to the exclusion principle and are called *bosons*. Photons, alpha particles and He atoms are bosons. All particles fall in one of these two categories. Particles are described mathematically as wave functions, in (relativistic) quantum mechanics. These wave functions are solutions of relativistic wave equations. The number of "components" of a wave function depends on the spin of the particle that describes. As an example, a spin-1 particle with mass is described by a triplet of wave functions: two for the polarisation states which are transverse (perpendicular) to the particle momentum and one for longitudinal polarisation (parallel to the momentum). The photon has spin 1, but has no mass and has no longitudinal polarisation. It is described by the two components of the electric field which are perpendicular to the photon propagation direction. In general bosons are described by an odd number of components, fermions by an even number of components.

Fermions and Bosons:

- spin $1/2, 3/2 \ldots$ = fermions ;
  Pauli's exclusion principle applies
- spin $0, 1, 2, \ldots$ = bosons;
  No exclusion principle, many particles can be in the same state.

6.6 Elementary Bosons

The elementary vector bosons that we know are all carriers of fundamental interactions:

Electromagnetic: photons have spin $1\hbar$ and mediate the electromagnetic interaction. These are massless force carriers of the electromagnetic interaction and are described mathematically by vector wave functions $\overset{\leftrightarrow}{A}$, with no transverse component. In quantum electro dynamics (QED) the photon is indicated with γ and is described by a 4-vector field

$$\mathbf{A(x)} = \epsilon A_0 \sin(\kappa x) \tag{6.40}$$

where ϵ is the polarisation 4-vector and κ is the propagation 4-vector, as shown in Fig. 6.8.

Weak: $W^{\pm}$ and Z^0 are also "vector" bosons, they have spin $1\hbar$, their non-relativistic wave function is a 3-component vector, and they are the carriers of the weak sector of the electroweak interaction. Their masses are m_W =

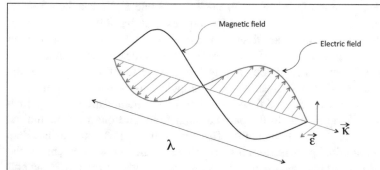

Fig. 6.8 Electromagnetic wave: the electric and magnetic field are perpendicular to each other; the propagation vector $\vec{\kappa}$ is along the propagation direction, the polarisation vector $\vec{\epsilon}$ is perpendicular to $\vec{\kappa}$

80.385 GeV/c^2 and $m_Z = 91.1876$ GeV/c^2. They were predicted theoretically in 1968 and were discovered experimentally at the Super Proton-antiproton Synchrotron (S$p\overline{p}$S) in 1983. The vector bosons decay almost immediately to a fermion–antifermion pair. As an example, Fig. 6.2 shows the invariant mass distribution for the Z^0.

> Sheldon Glashow, Steven Weinberg and Abdus Salam were awarded the Nobel prize in 1979 for the *electroweak* theory which predicted the $W^\pm$ and the Z^0; Carlo Rubbia and Simon Van Der Meer were awarded the Nobel prize in 1984 for the experimental discovery of the vector bosons.

Strong: Eight vector bosons called *gluons* mediate the strong nuclear force, which is described by the quantum chromo dynamics (QCD). They have spin $1\hbar$ and are massless. They have no electric charge, but have a strong charge, which is conventionally called *colour* and has nothing to do with the colours in our ordinary experience.

6.7 The Higgs Boson

The Higgs boson is the only fundamental particle which has no electric charge, spin zero, and has its own special place in the realm of particles. It is described by a one-component wave function, which does not change if we rotate the reference system. Its presence was postulated in 1964 by Peter Higgs and, independently, by Francois Englert and Robert Brout, as an elegant mathematical solution to provide mass to particles, while still preserving fundamental symmetries. The explicit mass terms are replaced with an interaction with the Higgs boson. The peculiarity is that the strength of this interaction is proportional to the mass of the particle. The Higgs boson has been detected in 2012 at the Large Hadron Collider, as a peak in the invariant mass of its decay products: two photons or a pair of Z^0 vector bosons. As the boson is electrically neutral, and photons are massless, the decay to two photons has an extremely low branching ratio, due to quantum corrections which are calculated using higher order calculations. The mass of the Higgs boson was not predicted by the theory and turned out to be

$$m_H = 125.09 \pm 0.24 \, \text{GeV}/c^2 \tag{6.41}$$

The coupling of the Higgs boson is peculiar, because it is proportional to the mass of particles: it has a very strong interaction with the *top* quark which is the heaviest known particle, a strong interaction with the $W^\pm$ and with the Z^0 and a very weak interaction with all others, as shown in Fig. 6.9. It has been discovered in proton–proton collisions with its decay to photons. As photons are massless, and the Higgs boson is electrically neutral, there is no direct coupling between these two

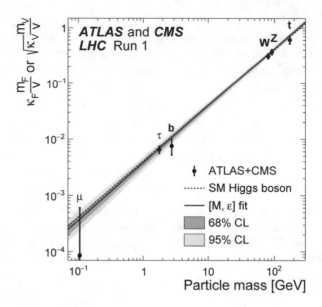

Fig. 6.9 The coupling of the Higgs boson to other particles is proportional to the mass of the particle. This has been measured in proton–proton collisions at the Large Hadron Collider (LHC) by two international collaborations ATLAS and CMS, using data collected at a centre of mass energy of 7 and 8 TeV ("Run1"). In the vertical axis is the dimension-less coupling strength, on the horizontal axis the particle mass. ATLAS and CMS Collaboration, J. High Energ. Phys. (2016) 45 Springer (2016)

particles. However some *virtual processes*, which will be mentioned later, involving a contribution from the top quark, make possible its coupling both with gluons, for production, and with photons for its decay. Although the branching fraction of the Higgs boson decaying to two photons is extremely low, this decay mode is the most distinctive and useful because of the relatively low *background* due to the production of pairs of high-energy photons, as shown in Fig. 6.10.

Peter Higgs and Francois Englert were awarded the Nobel prize in 2013 for the theoretical discovery of a mechanism that contributes to our understanding of the origin of mass of subatomic particles. Robert Brout (1928–2011) from Belgium also contributed to the theoretical prediction of the boson.

6.8   Elementary Fermions: Quarks and Leptons

The fundamental particles which make up the world in terms of matter (and anti-matter) are the *leptons* and *quarks*. They are fermions and are the building blocks

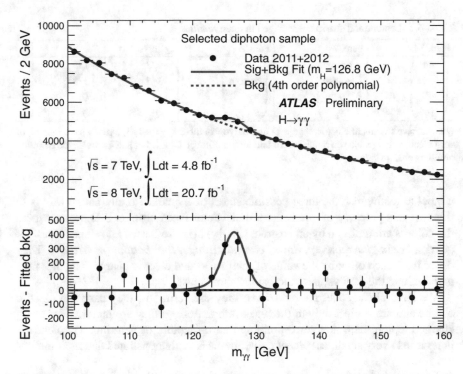

Fig. 6.10 Invariant mass distribution of two high-energy photons in "events" from *pp* collisions at a centre of mass energy of 7 and 8 TeV. The mass of the Higgs boson is measured from the invariant mass of the two γ produced in the decay $H \rightarrow \gamma\gamma$. The plot below shows the background-subtracted data, where the peak in invariant mass is more evident (from Phys. Lett. B 726 (2013) Elsevier (2013))

of ordinary matter as we know it, as well as the particles that are created by cosmic ray interactions and the matter that was present at the early stages of the universe.

> *Leptons* from the Greek λεπτος which means thin, light.
> The word quark originates from
>> *Three quarks for Muster Mark!*
>> *Sure he hasn't got much of a bark*
>> *And sure any he has it's all beside the mark.*
> From *the Finnegans wake*, by J. Joyce.

Ordinary matter is made of electrons and two types of quarks, which we call *up* and *down*. Electrons have an electric charge $-e$, while quarks have a fractional electric charge: $q_{up} = 2/3\,e$ and $q_{down} = -1/3\,e$. Electrons are subject only to electromagnetic and weak interactions, quarks are subject to all three interactions. They carry a "colour" charge for the strong interactions: each quark comes in three colours. This property has nothing to do with the colours of everyday life: it is just

Table 6.2 Leptons and quarks are arranged in three *families*

Fam. 1	Fam. 2	Fam. 3	$Q(|e|)$	B	L
up	charm	top	$+2/3$	$+1/3$	0
down	strange	bottom	$-1/3$	$+1/3$	0
e^-	μ^-	τ^-	-1	0	1
ν_e	ν_μ	ν_τ	0	0	1

Quarks have fractional electric charge Q and baryon number $B = +1/3$; *leptons* have integer, or zero electric charge, baryon number zero and lepton number $L = 1$. For anti-particles the quantum numbers change sign

a label to distinguish the three possible states of a quark. An alternative way is to think of this property of quarks as a unit vector in an abstract 3-D space, where the colours are replaced by an abstract $|x\rangle$, $|y\rangle$, $|z\rangle$. So far we know six "types" of quarks, or six "*flavours: up, down, charm, strange, top, bottom (or beauty)*". They differ from each other by the value of their mass and electric charge. The first two, u and d, and the quark s, have almost the same mass.

We have already met the *neutrinos*: they are fermions, they carry no electric charge and are subject only to the weak interaction, so they are leptons. For every charged lepton there is a corresponding neutral lepton, or neutrino. The neutrino mass is also very small, but recent experiments have demonstrated that their mass is not zero.

Leptons and quarks can be arranged in three "families", or "generations" as columns in Table 6.2, where the first three columns indicate particles of the three families, the last three columns indicate the corresponding quantum numbers.

In addition to fermions in Table 6.2 we know the following fundamental particles

$$\gamma, Z^0, W^\pm, g, H$$

which are mediators of interactions.

To be correct, we should have three "copies" of each quark, one for each "colour" charge. When considering the existence of colours, within each family the sum of the electric charge is zero:

$$Q = 3 \times 2/3 - 3 \times 1/3 - 1 = 0 \tag{6.42}$$

The main difference between families is the value of the particle masses: they increase from the first to the third family, as shown in Table 6.3. We don't know why there are three families: we know experimentally that there are no further families with the same structure, in particular we know that there are only three types of light neutrinos.

Table 6.3 Experimental values of the masses of fundamental fermions

	Mass (MeV)
Lepton	
e	$0.5109989461 \pm (31)$
μ	$105.6583745 \pm (24)$
τ	1776.86 ± 0.12
Quark	
u	2.2 ± 0.5
d	4.7 ± 0.5
c	1280 ± 30
s	96 ± 6
t	173100 ± 600
b	4180 ± 40

The experimental errors in parenthesis refer to the last two digits. Data from C. Patrignani et al. (Particle Data Group), Chin. Phys. C 40, 100001 (2016)

In Table 6.2 each symbol of a particle corresponds to a wave function. For instance, e^- really means $\psi_e(\mathbf{x})$, i.e. the wave function of an electron; similarly, for the quark "up" (u) and the quark "down" $d = \psi_d(\mathbf{x})$. The same applies to other fundamental particles. In more advanced books these wave functions are properly extended to represent spin $1/2$ particles, by using a multi-component wave function.

6.9 Anti-particles

All fundamental particles with spin $1/2$ have a corresponding anti-particle. The existence of the anti-electron was predicted by P.A.M. Dirac explicitly in 1931, as a mathematical necessity for a relativistically invariant description of particles with spin $1/2$. The anti-electron, or *positron* was discovered in cosmic rays in 1932, while 2 years later the positron-emitting β^+ decay was discovered. Anti-particles have exactly the same mass as the corresponding particles, but opposite electric charge. When a particle and its anti-particle "meet" they may, or may not, because of the probabilistic nature of quantum mechanics, produce a bound state and then decay into photons. As an example, we have the electron–positron bound state, called *positronium* which sometimes is indicated with *Ps*. Positronium is formed with low-energy positrons, e.g. those originated from a β^+ decay, and is produced in two "forms", spin-zero and spin-1, which decay into two and three photons, respectively. In high energy $e^+ e^-$ collisions the initial particle–anti-particle can transform themselves into a different pair of particle–anti-particle, like

$$e^+ e^- \to \mu^+ \mu^- \quad \text{or} \quad e^+ e^- \to q\bar{q} \tag{6.43}$$

Anti-particles are indicated with a bar above the symbol: in Eq. (6.43), q indicates a quark and $\bar{q}$ indicates an anti-quark; if ν_μ indicates a neutrino of the second family, which is called muon-neutrino, then the symbol $\bar{\nu}_\mu$ indicates the corresponding anti-neutrino. For charged leptons (e, μ, τ) it is customary to indicate their electric charge as a superscript: e^- for an electron and e^+ for a positron; by extension from the case of electrons, the negatively charged leptons are considered as particles, while the positively charged leptons are anti-particles. Particles and anti-particles appear and disappear in pairs: as an example, positronium decays into photons. The opposite process is called "pair creation" and it occurs when one (or in general more, to conserve energy-momentum) fundamental boson transform into a particle–anti-particle pair. As an example, if a photon traversing a medium has sufficient energy it can transform itself into a pair of e^+e^-. An exception to this occurs when the charged weak $W^\pm$ boson is involved: it does not couple to a quark and its own anti-quark, but to a different pair of quark–anti-quark flavours. Charged weak decays are therefore "flavour changing". A similar process to those in Eq. (6.43) is

---

Incidentally, the mathematical operator which transforms a particle quantum state, or wave function, into the corresponding anti-particle state is called "*charge conjugation*", and indicated with C:

$$C \, |e^-\rangle = |e^+\rangle \tag{6.44}$$

The electromagnetic and strong interactions are the same for particles and anti-particles, while the weak interaction does not conserve C and space inversion, which is also called *parity* P.

$$D^+ (= c\bar{d}) \rightarrow \mu^+ \, \nu_\mu \tag{6.45}$$

Here the D^+ indicates a *meson*, a bound state which will be described later.

6.10   More Quantum Numbers

We can now define some quantum numbers, which are useful to describe systems of particles. We have already mentioned in Table 6.2 the *baryon number*, which is $B = 1/3$ for quarks and $B = 0$ for lepton and the *lepton number* which is $L = 1$ for electrons, muons, taus and neutrinos, and is $L = 0$ for all quarks. These numbers are important because both are conserved by all interactions. We can also define the flavour-related leptonic number (which should be called the family related leptonic number). Not only the total number of leptons is conserved, but this holds true within each family. So we have the electronic number (or flavour), the muonic number and the tau-onic number, each of which is separately conserved. They are defined as

$$\mathcal{L}_e = N(e^-) + N(\nu_e) - N(e^+) - N(\overline{\nu}_e) \tag{6.46}$$

$$\mathcal{L}_\mu = N(\mu^-) + N(\nu_\mu) - N(\mu^+) - N(\overline{\nu}_\mu) \tag{6.47}$$

$$\mathcal{L}_\tau = N(\tau^-) + N(\nu_\tau) - N(\tau^+) - N(\overline{\nu}_\tau) \tag{6.48}$$

Each of them is individually conserved by all, but the weak interaction: neutrino oscillations, which have been recently discovered, indicate that the three equations above ((6.46)–(6.48)) are mostly valid, but have to be considered just as an approximation: neutrinos can change nature and "oscillate" between a type and another: $\nu_e \to \nu_\mu \to \nu_e$. So, while the total lepton number is conserved, the family lepton number is not. Experimentally there is no evidence of any flavour changing neutral process, of the type $\mu \to e + \gamma$, or $\mu \to e + e^+ e^-$.

> The possibility of neutrino oscillations was predicted by Bruno Pontecorvo in 1957. The 2015 Nobel prize was assigned to Takaaki Kajita and Arthur McDonald for the experimental discovery of neutrino oscillations, which indirectly shows that neutrinos have a small mass.

6.11 Feynman Diagrams

Recalling Eq. (6.9), a process, which can be a decay or a scattering, occurs with a probability that is factorised in a *phase space density* factor ρ_f, and a second term, which depends on the interaction and is called *transition amplitude* from the initial to the final state. The latter is indicated with $\mathcal{M}_{if}$ and is also called a *"matrix element"*. The formula for the transition amplitude, in the Born approximation, has a structure: it is a sum of terms of a series. Each term corresponds to a power of the coupling constant of the interaction: α, α_s, α_{EW} and can be translated into a graph, which indeed resembles the pictorial representation of a scattering process (Fig. 6.11). Each factor of a term corresponds to an element of the graph. This graph can be imagined to be in the (t, x) plane and is called a *Feynman diagram*. It is beyond the scope of this course to calculate the transition amplitudes $\mathcal{M}_{if}$ using Feynman diagrams; however, we can use them as a pictorial description of the reaction. These diagrams follow specific rules and conventions. Fermions are represented by continuous lines with arrows going forward in time, anti-fermions are represented by continuous lines with arrows going backward in time. The anti-fermions that are going back in time should not be taken literally: positrons are not coming from the future. This convention is useful to translate a graph into a formula and to correctly draw Feynman diagrams. The incoming and outgoing

The diagrams are named after Richard Feynman, 1918–1988, from the USA. He was awarded the Nobel prize in physics in 1965 for his fundamental work on quantum electrodynamics and particle physics.

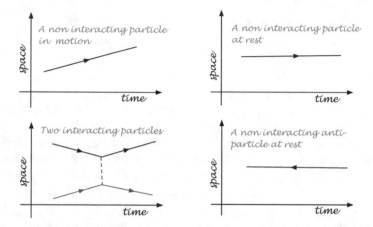

Fig. 6.11 The Feynman diagrams originate from the particle trajectory in the space-time plane. Non-interacting particles are represented by straight lines. Their slope cannot exceed the speed of light. Particles at rest are represented by horizontal lines. Graphs translate directly into formulas, so in Feynman diagrams only the topology of the graph is important, not the line slopes. Conventionally anti-particles are represented with an arrow pointing back in time. The convention can be explained with the CPT theorem: anti-particles, which are obtained by charge conjugation C, are particles in reverse time-space (PT) so they are represented with a reversed arrow and are said to go "backward in time"

lines are connected by "internal" lines, or *propagators*, which correspond to the interaction. If there are less than two internal lines, the diagram is said to be "at the tree level". The internal lines represent the force carriers, or "virtual" states of the particle. Some special graphs, which only involve force carriers, may have no internal line. Other conventions are shown in Fig. 6.12:

- Photons, γ, are represented by wavy lines.
- Vector bosons, $W^{\pm}$, Z^0, are represented by wavy lines, possibly of different style.
- Gluons, g, are represented by spring-shaped lines.

Vertices represent interactions. In general only three particle lines meet at a vertex: two for fermions and one for the force-carrying boson. In some special cases four force-carrying bosons can meet at a vertex: four gluons or 4 week bosons, but this is beyond the scope of this book. Electric charge is conserved at all vertices. The number of incoming (fermions minus anti-fermions) must equal the number of outgoing (fermions minus anti-fermions). Graphically, this means that there is a "fermionic flow" at each vertex, with no source or drain for fermions. Photons only interact with charged particles and consequently there is no vertex involving photons

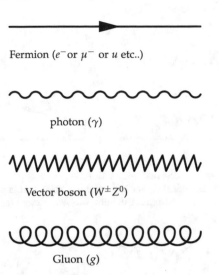

Fermion (e^- or μ^- or u etc..)

photon (γ)

Vector boson ($W^\pm Z^0$)

Gluon (g)

Fig. 6.12 Line convention to represent fermions, photons, vector bosons and gluons

and neutrinos. Gluons only interact with quarks and with other gluons, and therefore there is no vertex involving gluons and leptons. The Higgs boson is represented with a dashed line and only couples to massive particles: there is no vertex involving the Higgs boson and photons or gluons. As neutrinos have a very small mass, their coupling to the Higgs boson can be considered to be zero for most cases.

The incoming particle can be seen as emitting a quantum of the "preferred" force and starts scattering, the target particle absorbs a quantum and is scattered. It is important to keep in mind that the Feynman diagrams are a graphical visualisation of a formula to calculate $\mathcal{M}_{fi}$: e.g. for every "internal" line a "propagator" must be used in the formula. We have already seen its general form. The external lines, incoming and outgoing, translate into wave functions. The slopes of lines are not important for Feynman diagrams, what is important is the *topology*, i.e. how lines are connected: the same connections correspond to the same formula, as shown in Fig. 6.13.

The matrix element of a physical process is calculated by adding all Feynman diagrams which are not equivalent and have the same initial and final state. An example is shown in Fig. 6.14, where two diagrams contribute to the electron–positron annihilation to two gammas. The cross section is proportional to $|\mathcal{M}_{if}|^2$. When more than one sub-process contribute, with identical initial and final states, the matrix elements are added:

$$\mathcal{M}_{if} = \mathcal{M}_{if}^{(a)} + \mathcal{M}_{if}^{(b)} + \mathcal{M}_{if}^{(c)} \tag{6.49}$$

and the cross section is proportional to the square of the sum. As each term $\mathcal{M}_{if}^{(j)}$, ($j = a, b, c \ldots$) is not necessarily positive, interference effects are possible.

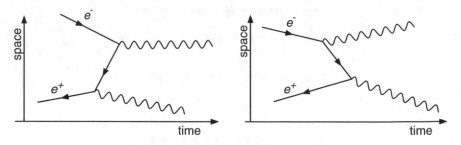

Fig. 6.13 These two diagrams describe with a different time order of the same process. They translate into identical mathematical formulas for the scattering matrix element. Their topology is the same, although the slope of the lines is different, this is irrelevant for Feynman diagrams

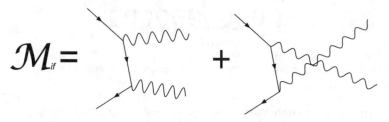

Fig. 6.14 At the lowest approximation level, where no *loop* line is considered, there are two topologically different Feynman diagrams for the process $e^+ e^- \rightarrow \gamma \gamma$. The corresponding formulas are two terms which are added to form the matrix element

Each of the terms $\mathcal{M}_{if}^{(j)}$ is in reality a sum of *amplitudes* $\mathcal{A}$ times powers of the coupling constant of the interaction:

$$\mathcal{M}_{if}^{(a)} = \sum_{n}^{\infty} \alpha^n \mathcal{A}_n^{(a)} \tag{6.50}$$

The number of terms is infinite, but for all practical calculations truncated sums are used, up to a given power of the coupling, or *order*. When the coupling is small the result is satisfactory already at the lowest order (LO) $\mathcal{M}_{if\,(\text{LO})}^{(a)}$.

6.12   Feynman Diagrams: Examples

The two Feynman diagrams corresponding to the Compton scattering are shown in Fig. 6.15. In Fig. 6.16 the main vertices for electro weak interactions are shown. In Fig. 6.17 two important Feynman diagrams are shown, for the two beta decays β^+ and β^-. The two processes have about the same Feynman diagram, with the same vertices, so their matrix element $|\mathcal{M}_{if}|^2$ also has the same value. The two other quarks don't take part in the process, and are called "spectators". Neutrons decay into protons because the neutron mass is (slightly) larger than the proton mass. This

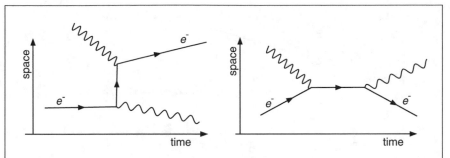

Fig. 6.15 Feynman diagram of Compton scattering. The Klein–Nishina cross section (Eq. (5.34)) can be calculated using this diagram

makes some quantum states are available to the final states, making the factor ρ_f larger than zero. Conversely, the proton mass is lower than the neutron mass and, for isolated protons, no final state is available for a decay into neutrons ($\rho_f = 0$). We'll see in the next chapter that this may no longer be the case if we consider the mass energy of the whole nucleus, and we observe that also the β^+ decay occurs in nature.

In Fig. 6.18 a *higher order* process is shown, with the only purpose of mentioning that more complicated diagrams correspond to either rare processes, as the one in the picture, where the Higgs boson couples, indirectly to gluons and to photons, or to "better" approximation of processes which also occur at tree level. In this case they are additional terms of a series which converges to the result, which is typically a cross section (Fig. 6.19).

6.13 Composite Particles: Mesons and Baryons

The nuclear strong interaction holds together the *quarks* to form

mesons   bound states of one quark and one anti-quark;
baryons   bound states of three quarks.

Baryons and mesons (Figs. 6.20 and 6.21) are collectively called *hadrons* and are all subject to the strong nuclear interaction.

Neutrons and protons are baryons:

- **neutron** = (up, down, down);
  electric charge $q_n = 2/3 - 1/3 - 1/3 = 0$
- **proton** = (up, up, down)
  electric charge $q_p = 2/3, 2/3 - 1/3 = +1$

Many other particles were discovered, initially studying the composition of cosmic rays, then with particle accelerators, with a large number discovered in the

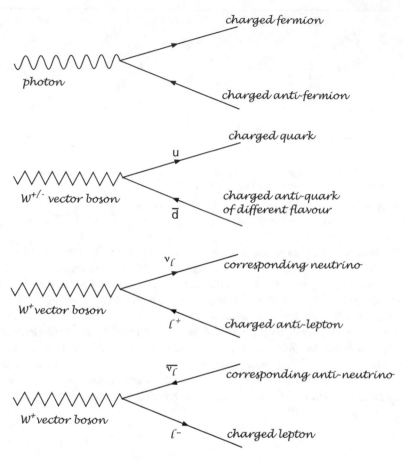

Fig. 6.16 Vertices of Feynman diagrams for some fundamental forces. These are only parts of Feynman diagrams used to calculate a physical process. Conservation rules, like charge and spin conservation, apply at each vertex. The $W^{\pm}$ boson couples with leptons and anti-leptons of the same family, and thus conserves the lepton flavour number, but it can couple to quarks of different families. Examples: $W^{+} \rightarrow u\bar{d}$; $W^{+} \rightarrow u\bar{s}$; $W^{+} \rightarrow e^{+}\nu_{e}$; $W^{+} \rightarrow \mu^{+}\nu_{\mu}$; $W^{+} \rightarrow \tau^{+}\nu_{\tau}$; Other vertices can be constructed this way, provided charge and lepton number are conserved. As an example by charge conjugation of the last one we obtain $W^{-} \rightarrow \tau^{-}\bar{\nu}_{\tau}$; These diagrams can be "rotated" at will and stuck together to form diagrams corresponding to physics processes

years 1960–1970. The classification of these, in terms of quantum numbers and regularity patterns, is based on the quark model, quantum mechanics and group theory. It was realised that some of the new particles, with extremely short lifetime, were orbital excited states of other particles and are called "resonances". They still have their own mass, but the quark content is the same as other particles. Using the language of atomic physics, where the energy levels of atomic electrons are studied, the subject is called *hadron spectroscopy*.

We give below a short and a longer introduction to hadron spectroscopy.

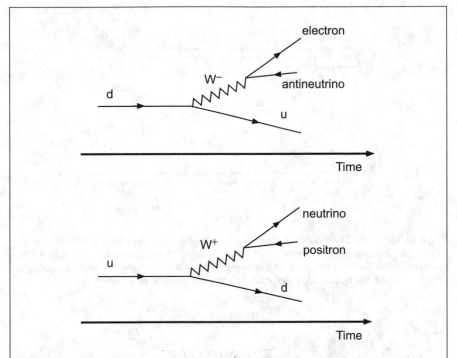

Fig. 6.17 The fundamental diagrams for the β^+ and β^- decays, where we have used the last two vertices of the figure above. Conservation rules apply at each vertex. In this case $W^{\pm}$ couples with leptons of the same family on one end and with quarks of the same family on the other end

6.14   Hadron Spectroscopy: A Short Introduction

Change of notation: so far we have indicated the spin with s. To avoid confusion with the strange quark, from now on the spin is indicated with J.

Hadrons are composite particles which are made of quarks and are held together by gluons. We use their quantum numbers to classify them. Primarily, the baryon number separates *mesons*, which are bound states of a quark and an anti-quark, and therefore have baryon number $B = 0$, from *baryons*, which are bound states of three quarks and, therefore, have $B = 1$. Within each category, the second most important quantum number is the *spin*. In general, mesons are composite *bosons* with spin $J = 0, 1, 2 \ldots$, while baryons are composite fermions, $J = 1/2, 3/2, \ldots$. The intrinsic *Parity* is another useful quantum number, but its description is left to a more advanced course. Given this triplet of quantum numbers, $|B, J, P\rangle$, the composite particles can be further grouped into multiplets: particles of the same multiplet behave equivalently under the strong interaction. For instance, the mesons with $|B, J, P\rangle = |0, 0, -\rangle$ form two sets: a single particle (η'), called a *singlet*,

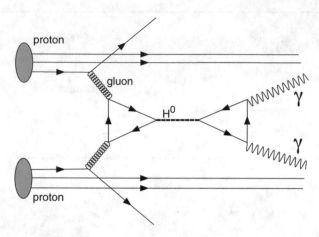

Fig. 6.18 One of the Feynman diagrams of Higgs production in $p-p$ collisions, with H^0 decaying to $\gamma\gamma$. This is a higher order process, with triangle shaped loops. The electrically neutral Higgs boson couples indirectly to photons via this *loop* process, where the *top quark* provides the highest contribution. An analogous term couples the colourless Higgs boson to gluons

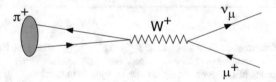

Fig. 6.19 Weak decay of a composite particle, a charged π-meson, which is a bound state of quark–anti-quark of the first family. $\pi^- \to \mu^- \bar{\nu}_\mu$ The charged lepton and the corresponding anti-neutrino are from the same family. The charged lepton has the same charge as the meson

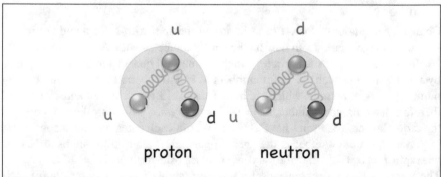

Fig. 6.20 Quark composition scheme of the proton (left) and the neutron (right)

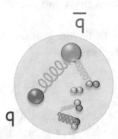

Fig. 6.21 An artist's view of the quark composition of a meson. Gluons are shown as little springs. Besides the component quark and anti-quark pair, other *virtual* quark–anti-quark pairs are continuously created and destroyed

and a set of eight particles which behave equivalently if we ideally turn off all but the strong interaction: $\pi^+, \pi^-, \pi^0, \eta^0, K^0, K^-, K^+, \overline{K}^0$. In other words, for the strong interaction we only have two particles: the η' and any of the particles in the octet. This information should be sufficient for an introductory course, but for completeness we can go more in detail in the following sections.

6.15 Meson Classification

It is important to make clear that the classification of hadrons is not driven by a stamp-collecting passion, but is a fundamental key to understand the strong interaction: particles which are in the same "category", although are distinguishable for other features, behave essentially as the same particle for what strong interactions are concerned, in particular in terms of scattering amplitudes. Most of the classification work was initiated before the quark model emerged and was essential to build this model. In the light of the quark model the hadron characterisation is much clearer. Here we'll use the opposite approach and will start from quarks to form bound states. Let's start with mesons, which are formed with a quark and an anti-quark. By definition their baryon number is zero. First of all we have to choose the total spin and angular momentum: we can form scalar mesons, which have zero spin, or vector mesons, with spin-1.

$$(1/2, 1/2)_{\text{spin}} \Rightarrow \begin{cases} J = 1 \text{ triplet state } J_3 = -1; 0; +1 \\ J = 0 \text{ singlet state } \quad J_3 = 0 \end{cases} \tag{6.51}$$

Other excited states, with higher orbital angular momentum, are also possible, giving rise to a large number of so-called meson *resonances*, each behaving as a short-lived particle, with its own mass and quantum numbers. Let's focus on the scalar mesons.[3]

We have five different types, or flavours, of quarks which form bound states, so the total number of *pseudoscalar meson* is

$$N_{p.s.} = 5^2 = 25$$

While this result is correct, at least for mesons, the way quarks are arranged is not as simple: for mesons formed with u, d, s quarks the quark content of the mesons is not pure, but is a linear combination of quark–anti-quark states. Group theory is needed for a mathematical description of this classification. To understand this we should imagine turning off the electric charge and assigning all quarks the same mass value. This way the only label which identifies a quark is its *flavour*: u, d, s, c, b. In an abstract way we can think of it as a five-dimensional vector space, and transformations in this space are elements of the SU(5) group. As the mass of the b-quark is much larger than the mass of the others, this is really a broken symmetry, and for c-quark mesons the symmetry is only approximate. Let's start with only the first two quarks, u and d. Their mass is indeed very similar and when turning off the electromagnetic interaction, they really behave almost as two states of the same particle. These two states can be described by a vector space where SU(2) transformations operate. This is the same group which operates on the spin, and this property is called *isotopic spin*, or *isospin*: having turned off the electromagnetic interaction, there is only one quark with isospin $1/2$. In this picture the quark *up* has "third component" of the isospin $I_3 = +1/2$, or "up", while the quark *down* has $I_3 = -1/2$, or "down". This way, combining a quark with an anti-quark becomes mathematically similar to combining their spins, and we know from quantum mechanics, or from group theory, that combining spins $(1/2, 1/2)$ just as in Eq. (6.51) we obtain:

$$(1/2, 1/2)_{isospin} \Rightarrow \begin{cases} I = 1 \text{ triplet state } I_3 = -1; 0; +1 \\ I = 0 \text{ singlet state } \quad I_3 = 0 \end{cases} \tag{6.52}$$

This is shown graphically in Fig. 6.22, in what is called a *weight diagram* in group theory language. These states are identified as the mesons π^-, π^0, π^+ (which are called *pions*) for the triplet and η as the singlet state, in the first approximation. It should be noted that the third component of isospin has nothing to do with the z direction in ordinary space, but it is a mathematical abstraction. The strong interaction depends on the isospin of the system, not on its third component. This statement can be verified experimentally as we are able to produce secondary beams of pions with accelerators.

---

[3]These scalar mesons are actually *pseudo*scalars, because their wave function changes sign under *parity*. Parity P is the transformation that changes sign to the space coordinates: $P : \vec{x} \rightarrow -\vec{x}$.

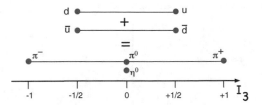

Fig. 6.22 A quark (u/d) and an anti-quark $(\overline{u}/\overline{d})$ are two states of isospin $1/2$. By combining them we obtain a meson of isospin 1, which has three possible states of I_3, corresponding each to a meson (pion), and a meson of isospin $I = 0$ which is only one state, a singlet state, which we temporarily identify with the η meson. In group theory notation: $\mathbf{2} \otimes \overline{\mathbf{2}} \Rightarrow \mathbf{3} \oplus \mathbf{1}$ (SU(2))

The key point is that in combining quarks to form mesons we cannot simply consider quarks as marbles. The total number of possible mesons is the same as if the quarks were classical objects. However , the existence of isospin symmetry for quarks and the requirement that the resulting mesons are also eigenstates of isospin "rotations" results in a grouping of mesons according to their isospin value. In addition, the meson states are linear combinations of the classical pairs $(q_1, \overline{q}_2)$.

We are now ready to add a third type, or *flavour* of quark to the picture, the *strange* quark s. The vector space now has three dimensions, and the fundamental transformations in this space are operators of the SU(3) group. While the eigenstates of SU(2) are labeled by just one value, which can be $\pm 1/2$ in the **2**-dimensional space, the eigenstates of SU(3) are identified by a set of two values, one of them is the third component of the isospin, as in SU(2), the other is, in first approximation, the *strangeness*, a quantum number which is $S = -1$ for mesons containing a strange quark and $S = +1$ if a strange *anti*-quark is present. So the s quark itself has no isospin and strangeness $S = -1$, while the anti-quark $\overline{s}$ has strangeness $S = +1$, as shown in Fig. 6.23. To form mesons with three flavours of quark and anti-quark we use the same method as we have done with two flavours. This time, however, we have to combine the states of the 3-dimensional representations of SU(3). We have to use a result from group theory. Its proof is outside the scope of this book. We obtain just two different particles for the strong interaction: a *singlet* state, with no isospin and no strangeness, and an octet of states, which are all equivalent from the point of view of the strong interaction and are labeled by two numbers:

the strangeness and the third component of isospin (s, I_3), as shown in Fig. 6.24. We identify the singlet state with the meson called η'. The eight particles in the octet are $\pi^+, \pi^-, \pi^0, \eta^0, K^0, K^-, K^+, \overline{K}^0$. The physical particles η^0 and η' are a mixture of the two, but this is a complication that goes beyond the scope of this introductory book. In group theory notation we write

$$\mathbf{3} \otimes \overline{\mathbf{3}} \Rightarrow \mathbf{1} \oplus \mathbf{8}; \quad \text{(SU(3))} \tag{6.53}$$

with respect to the spin the group is different. We are doing with groups all possible combinations of quark–anti-quark, rotating each of them independently of the other

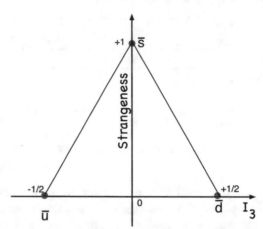

Fig. 6.23 The eigenstates of SU(3) can be labeled with two indices. In the approximate SU(3) flavour symmetry "the" quark assumes different names according to these indices. They can be chosen as the strangeness and the third component of the isospin. The vertices are the eigenvalues of the anti-quark flavour states, with respect to these two operators, in the 3×3 conjugate representation, which in group theory is indicated as $\overline{3}$. In other words, the anti-quark state $\overline{s}$ has isospin $I_3 = 0$ and strangeness $S = +1$, the anti-quark state $\overline{u}$ has isospin $I_3 = -1/2$ and strangeness $S = 0$

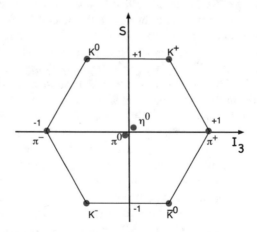

Fig. 6.24 The octet of pseudoscalar mesons is represented in the isospin (x) and strangeness (y) plane, according to the eigenvalues of the **8**-dimensional representation of SU(3)

in the "flavour vector space". Therefore we use the symbol $\otimes$. The result are two independent sets of states, i.e. particles, each being an eigenstate of a representation of SU(3). The next step is now quite clear: we introduce a fourth quark flavour, the *charm* c, and we make use of SU(4) operators, requiring that the physical states are eigenstates of a representation $\mathbf{4} \otimes \overline{\mathbf{4}}$-dimensional of SU(4). We introduce a *charm* quantum number, which is positive for the charmed quark and negative for the charmed anti-quark. The SU(4) eigenstates are identified by three numbers, which we can chose to be I_3, s, c and the corresponding graphical visualisation will be 3-dimensional, accommodating all $4^2 = 16$ mesons which we can form with four quarks and four anti-quarks. However, the mass of the *charm* quark is $m_c \approx 1.3$ GeV/c^2, which is much larger than the mass of the previous three quarks; even turning off all other interactions c-quarks still present their own identity. This is even more evident with the b-quark, with a mass $m_b \approx 4.1$ GeV/c^2, so it is not very useful to introduce SU(5). The next step is to combine three quarks to form the bound states which are called *baryons*.

The quantum number "strangeness" was introduced in 1953 by Murray Gell-Mann, who was awarded the Nobel prize in 1969 for the classification of elementary particles and their interactions. He coined the name for *quarks* and he is proponent, among other things of a mechanism for the neutrino to acquire a mass.

6.16   Baryon Classification

Baryons are bound states of three quarks, and to classify them we follow the same lines as we did for mesons, first restricting to the two lightest flavours, then adding flavours one by one. There is just a slight complication in the definition of the quantum numbers, and we'll have to use additional symmetry considerations, which restrict the number of possible states. As we did for mesons, first of all we have a look at the spin. A bound state of three particles with spin $1/2$ can have either all spins aligned, to form a particle of spin $3/2$, or just two spins aligned and the third in the opposite direction and form a system of spin $1/2$. Using the mathematical language of groups this is translated in the following way: the initial state of free particles is represented by three bi-dimensional subspaces, where the spin operators are the *direct sum* of three two-dimensional representations of SU(2). When forming a bound state the resulting particle is described by a spin space corresponding to one of the so-called *irreducible* representations of SU(2) in the available number of dimensions: in this case either a four-dimensional space, corresponding to a spin $3/2$, or a two-dimensional space, corresponding to spin $1/2$. The dimension of a representation is indicated as a boldface number:

$$\mathbf{2} \oplus \mathbf{2} \oplus \mathbf{2} \Rightarrow \mathbf{4} \oplus \mathbf{2} \quad \text{SU(2)(spin)} . \tag{6.54}$$

In actual fact, there are two possible sets of states which describe spin-$1/2$: one which is symmetric under interchange of quarks 1 and 2, and one which is antisymmetric for the same exchange. A linear combination of the two is going to be used. Let's focus on the spin $1/2$ states and combine the isospins of the three particles by "rotating" them independently of each other in the (u, d) isospin space. We form a *direct product* of three 2-dimensional representations of SU(2). We then find the irreducible representations of this product

$$2 \otimes 2 \otimes 2 \Rightarrow 4 \oplus 2 \oplus 2 \quad \text{(SU(2)isospin)} \tag{6.55}$$

The physical particles are eigenstates of the isospin I_3 operator representations. Also in this case we have two sets of states corresponding to isospin $1/2$: one which is symmetric under interchange of quarks 1 and 2, and one set which is antisymmetric. At this point some symmetry considerations are in order. Sets of indistinguishable particles with spin $1/2$ or $3/2$ are described by a global wave function, which is antisymmetric under interchange of any two particle, as required by Fermi-Dirac statistics (Eq. (6.38)). While studying baryons we have ideally "turned off" all interactions but the strong one. We have also neglected any possible mass difference between quarks, so the three quarks in a baryon are treated as indistinguishable. The total wave function is factorised in four parts:

$$\psi(q_1, q_2, q_3) = \phi_{\text{QCD}} \, \rho_{\text{L}} \, \eta_{\text{J}} \, \chi_{\text{iso}} \,. \tag{6.56}$$

Each factor depends on the three quark states (q_1, q_2, q_3): the first component is due to the strong interaction, and we know from arguments which we'll see later, is antisymmetric; the second factor ρ_{L} depends on the orbital angular momentum L, which we assume is zero, and therefore this wave function is symmetric; the third and fourth components describe the spin and the isospin, and their product must be symmetric. This is obtained by a linear combination of the two set of spin symmetric and isospin symmetric states $(1 \leftrightarrow 2)$ and spin anti-symmetric and isospin anti-symmetric $(1 \leftrightarrow 2)$. Therefore, the spin-$1/2$ isospin-$1/2$ baryon has two states, which we indicate with $|u, u, d\rangle$ and $|d, d, u\rangle$ for simplicity, but they have a more complicated structure. The two states, which are eigenvalues of the isospin third component, correspond to the well-known proton and neutron. They behave, from the point of view of the strong interaction, as the same particle, which is called *nucleon*. Protons have isospin third component $I_z = +1/2$, neutrons have $I_z = -1/2$.

If we try to combine spin-$1/2$ with isospin-$3/2$, or vice versa, the same symmetry arguments apply. The eigenstates corresponding to spin-$3/2$ are symmetric, and therefore they cannot be combined with mixed symmetry isospin states. So there is no baryon with $|J, I\rangle = |3/2, 1/2\rangle$ or $|J, I\rangle = |1/2, 3/2\rangle$. The only other combination left is $|J, I\rangle = |3/2, 3/2\rangle$. In this case both the spin and isospin states are symmetric, we have four baryons, corresponding to the four possible eigenstates of the third component of isospin: the "Delta resonances": $\Delta^-, \Delta^0, \Delta^+, \Delta^{++}$ (Fig. 6.25). They behave as the same particle for strong interactions, as is confirmed experimentally.

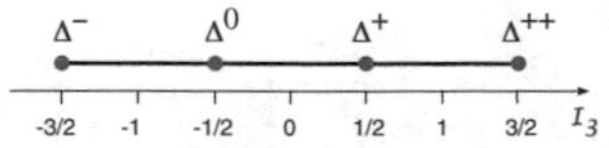

Fig. 6.25 The Δ resonances have spin $3/2$ and isospin $3/2$. Note the presence of a baryon with a positive electric charge $Q = 2|e|$, the Δ^{++}, which is composed by three u quarks

The existence of the states Δ^{++} and Δ^{-} has played a key role in discovering the symmetry that gave rise to our present understanding of the strong interaction, the so-called *colour* charge.

Having discovered the six possible baryons which are made with u and d quarks, we are now ready to add the *strange* quark s. As we have done with mesons, we now use SU(3) for the isospin part, while keeping the eigenstates of SU(2) representation for the spin. The irreducible representations of SU(3) are:

$$\mathbf{3} \otimes \mathbf{3} \otimes \mathbf{3} \Rightarrow \mathbf{1} \oplus \mathbf{8} \oplus \mathbf{8} \oplus \mathbf{10} \quad \text{(SU(3) flavour)} . \tag{6.57}$$

We therefore expect to be able to form 27 particles for any spin state, arranged in a singlet, two octets and a decuplet. However, once again symmetry considerations restrict the number of particle states to just one decuplet and one octet. We have to couple these states with the same spin states, just as in the case of two flavours. The SU(3) singlet eigenstate is antisymmetric by exchange of any two quarks, and there is no spin state that is also antisymmetric, to form a symmetric state. The same holds true for the eigenstates of one of the octet representations. The decuplet state is symmetric and can only be paired to the symmetric spin-$3/2$ states. The mixed symmetry octet states can be combined with the mixed symmetry spin-$1/2$ states, to give a symmetric wave function: we have the *baryon octet* and the *baryon decuplet*. As for the mesons, states are labeled by two indices: the third component of isospin I_3 and the strangeness S. To complicate slightly we can introduce the quantum number flavour *hypercharge* (Y), which is defined as

$$Y = B + S , \tag{6.58}$$

where B is the Baryon number, $B = 1$ for baryons and $B = -1$ for anti-baryons, and S is the strangeness, defined as $S = -1$ for every *strange* quark in the bound state. I_3 and Y can be used to label the baryon particle eigenstates, as shown in Fig. 6.26. The Gell-Mann Nishijima Nakano formula links the electric charge of baryons to the flavour hypercharge and the third component of the isospin.

$$Q = I_z + Y/2 = I_z + (B + S)/2 \tag{6.59}$$

Eq. (6.59) becomes clear within the quark model, assuming the strange quark has electric charge $Q(s) = -1/3$.

Adding more flavours, *charm* and *bottom/beauty* we can use, just as for mesons, higher symmetries and form charmed and beauty baryons. It is now time to explain the force that holds together quarks within hadrons.

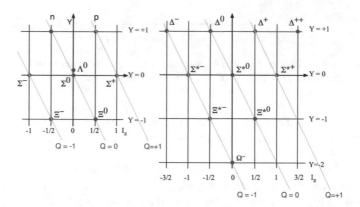

Fig. 6.26 The baryon octet (left) is a set of spin-1/2 particles which behave as the same particle with respect to the strong interaction. The particles are eigenstates of SU(3)$_{\text{flav.}}$ operators and are labeled with two sets of eigenvalues: the flavour hypercharge Y on the vertical axis and the third component of isospin in the horizontal axis. The Λ^0 and the Σ^0 have the same labels. Also the baryon decuplet (right) behaves as one particle when only the strong interaction is considered. The particles of this set have spin 3/2. Note the * to the Σ^* and Ξ^* decuplet states, indicating that they have the same quantum numbers as the corresponding octet particles, but have a higher spin. The Gell-Mann Nishijima Nakano formula for the electric charge is also illustrated with diagonal lines

6.17 The Strong Interaction

The strong interaction binds together quarks to form mesons and baryons. At a different energy scale level it also binds together nucleons inside atomic nuclei. Using an analogy, it is like the electromagnetic force binding together nuclei and electrons, but also a *residual* electromagnetic interaction among neutral atoms binds them together to form molecules.

We have given for granted the existence of quarks of fractional electric charge. One thing that we notice is that we have no experience of fractional electric charge, so whatever force binds quarks together must be so strong that it is impossible to separate them. This is normally called *confinement* of quarks. Also, we have no experience of nuclear force in everyday life, so it is a short-range force. This can be achieved by using massive "messengers" of the force; however, as we have seen, this also makes the force weak. An alternative solution requires that also the interaction quanta are *confined* to a neighbourhood of hadrons, and this is what happens for the strong interaction.

We have two well-defined energy ranges where we can describe this force: one is the static model, where we can imagine a potential around quarks, and we can try to calculate the energy levels, which correspond to the masses of the simplest systems of quark–anti-quarks, the mesons. The other energy range is obtained when we collide mesons and baryons among themselves or with electrons, at high energy.

The theory of strong interaction has to reconcile in one consistent description the experimental evidence that originates from

1. the energy levels of mesons and baryons,
2. the static model of mesons and baryons, as described in the previous sections,
3. the production of hadrons in e^+e^- collisions, as shown in Fig. 6.28,
4. the scattering of electrons and neutrinos on nucleons.

The energy levels of mesons are easier in two cases: when the mass of one of the quarks is much larger than the other, as in the case of b-flavoured mesons, like $B^\pm$, B^0 and B_s, when the quark–anti-quark pair have the same flavour and, therefore, the same mass. We call these states *quarkonia*: as an example the particle J/ψ is a bound state of $c\bar{c}$ quarks and the Υ is a bound state of $b\bar{b}$. We have mentioned before *positronium*, a bound state of an e^+ and an e^-, which is atom-like, in the sense that it is a neutral, electromagnetic bound state; quark–anti-quark bound states are held together by the strong force. By comparing the energy levels of the quarkonia with those of positronium, the potential of strong interactions can be inferred, with a term which is similar to the electrostatic potential $1/r$ and a second term which allows for quark confinement and is therefore proportional to the distance:

$$V_{\text{strong}} = -\frac{4}{3}\hbar c \frac{\alpha_s}{r} + kr \qquad (6.60)$$

The "constant" α_s is dimension-less and is characteristic of the strong interaction. Its value is only constant to first approximation, and ranges $0.1 \leq \alpha_s \leq 0.5$. The equivalent constant for the electromagnetic interaction has a value $\alpha_{\text{EM}} \approx 1/137$ (Eq. (5.37)), so the strong interaction is intrinsically strong. The potential as a function of radius r (Eq. (6.60)) is shown in Fig. 6.27. The origin of the factor $4/3$ will become clear later.

The b-flavoured mesons quark content:

$$B^- = (b\bar{u}), \; B^+ = (\bar{b}u)$$

$$B^0 = (\bar{b}d), \; B_s = (\bar{b}s)$$

$$\overline{B}^0 = (b\bar{d}), \; \overline{B}_s = (b\bar{s})$$

The linear term in the potential describes the *confinement* of quarks inside hadrons: it is not possible to separate quarks.

In the meson classification three particle the $\Delta^-(ddd)$, $\Delta^{++}(uuu)$ and the $\Omega^-(sss)$ play a special role, because their existence has led to the theory of strong interactions. In these examples of quark systems the Pauli's exclusion principle would be apparently violated: we have three quarks of the same flavour,

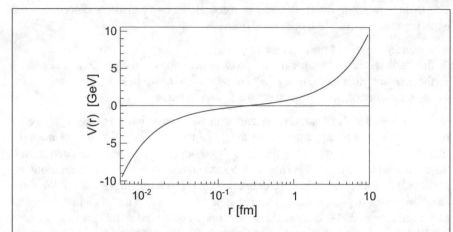

Fig. 6.27 The QCD potential, as a function of the distance between a quark and an anti-quark, as calculated from Eq: (6.60) with a constant value of $\alpha_s = 0.2$ and $k = 1$ GeV/fm. Note the logarithmic scale in the horizontal axis

indistinguishable and occupy the same state. So solve this problem a "strong charge" was introduced: each quark carries a strong "charge" which makes them different from one another. This charge has to comply with general consistency rules, like anti-quarks must carry the opposite charge. So it has to have a different structure from the electric charge: we need three different charges for quarks and the corresponding anti-charges for anti-quarks. The charges must also have all the same strength, so we can't call them (1), (2), (3), but letters may be more adequate: (a), (b), (c) or why not (r), (g), (b) like the so-called primary colours for the human eye. From this very indirect analogy the theory of strong interaction was named quantum chromo dynamics or QCD. We can combine systems of more than one quark, each with one of three "colour" charges, and we use exactly the same method as in the static model of mesons and baryons $(SU(3)_{\text{flavour}})$, with one additional rule: the combinations that correspond to observed particles are always an eigenfunction of a *singlet* representation of $SU(3)_{\text{colour}}$. Singlet eigenstates are antisymmetric with respect to an exchange of any two quarks, so the interaction wave function ϕ_{QCD} (Eq. (6.56)) is always antisymmetric. The three charge states are represented in a three-dimensional vector space where all vectors are unitary. In such a mathematical space we have exactly eight independent, non-trivial operators which transform one state into a different one. These operators form a group, which is $SU(3)$, the same group as the one we have used already to classify hadrons. However, instead of transforming a quark flavour into another flavour, like $u \rightarrow s$ (remember that we have ideally turned off the electromagnetic interaction), now these operators transform a quark strong charge into another strong charge, or in a more colourful language, they transform a quark colour into another colour. We identify these mathematical operators in the abstract space as the carriers of the strong force, the *gluons*. In QCD they have spin-1 and are massless.

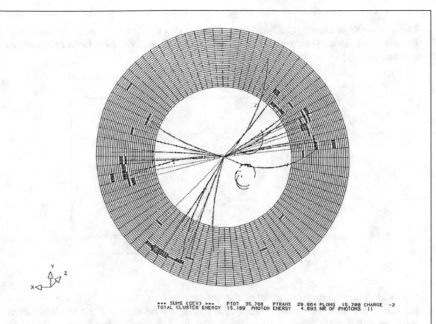

Fig. 6.28 In e^+e^- collisions at high energies $q\bar{q}$ pairs can be produced. Quarks appear as *jets* of particles. In some of these events a high-energy gluon can radiated by one of the quarks, producing an additional jet of particles: these events have 3-jets, as the one shown above: $e^+ + e^- \rightarrow q + \bar{q} + g$. Here the electrons come from above, perpendicular to the sheet, the e^+ beam from below, the curved tracks are the trajectories of the charged particles produced in the collision. They are bent by a magnetic field. A 3jet event from e^+e^- collisions at 31 GeV (JADE detector at PETRA accelerator, DESY, Germany, 1979). From P. Soeding, Eur. Phys. J. H 35, 3–28 (2010) Copyright Springer (2010)

The experimental evidence of the strongly charged gluons comes from e^+e^- collisions at high energy, where the kinetic energy is more than enough to produce a quark–anti-quark pair with high kinetic energy:

$$e^+ e^- \rightarrow q\bar{q} \rightarrow \text{jet, jet}. \tag{6.61}$$

While the quarks separate, the linear part, spring-like, of the strong potential (Eq. (6.60)) comes to play: eventually enough energy is stored in the spring tension to "break the spring" and generate more pairs of quark and anti-quark, at each new end of the spring; they ultimately combine and form mostly mesons. This process, which is called *quark fragmentation*, results in a focused shower, or *jet* of particles, which are all produced in a cone around the direction of the initial quark. In most cases we only observe two jets of particles, corresponding to the original quark-anti-quark pair; in some cases a third jet is observed, as shown in Fig. 6.28, which is an indication of emission of strong radiation: this jet corresponds to a high-energy

The *asymptotic freedom* of QCD was formulated by Frank Wilczek, David Gross and David Politzer, USA, 1973. They were awarded the Nobel prize in 2004. Also Gerard t'Hooft had contributed as a precursor on the same subject.

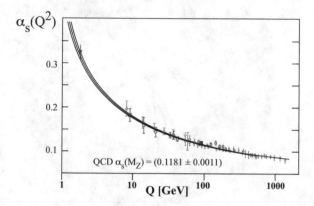

Fig. 6.29 The experimental value of the strong coupling "constant" α_s decreases as a function of the transferred momentum Q. Reprinted with permission from M. Tanabashi et al. (Review of Particle Physics), Phys. Rev. D, Vol. 98–1, p.155 (2018). Copyright (2018) by the American Physical Society

gluon being emitted by one of the two quarks:

$$e^+ e^- \rightarrow q \bar{q} g \rightarrow \text{jet, jet, jet} . \tag{6.62}$$

The emission rate is calculated exactly by QCD.

The last point, but also one of the most important, is the supporting evidence of QCD by scattering of high-energy leptons by nucleons. Apart from supporting the evidence of physical existence of quarks, these experiments show that the coupling constant α_s is not really a constant, but decreases when the momentum transferred by the scattering process increases, as shown in Fig. 6.29. From the practical point of view, quarks inside protons and neutrons behave more and more as free particles when probed by electrons and neutrinos at higher and higher energies. This effect is predicted by QCD and is called *asymptotic freedom*. The consequences of this running constant are important. At energies lower than about 1 GeV the value of the coupling is $\alpha_s \approx 1$ and it is not possible to use perturbation theory. Other approaches, like lattice calculations, can be used, or models, like the "bag model" of the proton. However, at high energies, when the momentum transferred is large with respect to the relevant scale of the order of the proton mass, we can use perturbation theory and Feynman diagrams, like we do at all energies for the other interactions.

The main consequence for this intrinsic difficulty of dealing with the strong interaction at low energy is that in Nuclear Physics we are limited to a phenomenological description of bound states of many nucleons, which results in several models for the atomic nucleus, as we'll see in the next chapter.

6.18 Problems

6.1 Calculate the ratio between the electromagnetic and the gravitational force between a proton and an electron.

6.2 Write all the possible Feynman diagrams at tree level corresponding to a decay of a Z^0 vector boson into fermion–antifermion pairs, taking into account that flavour changing decays of this particle are not present. The mass of the Z^0 is 91.1876 ± 0.0021 GeV/c^2, the mass of the top quark is 174 GeV/c^2. Comment on a possible decay of Z^0 to $t\bar{t}$. Neglecting all fermion masses with respect to the mass of Z^0, we can assume to first approximation that the phase space density is the same for all decays. As also the matrix element is the same, calculate the branching fraction of Z^0 to $c\bar{c}$ and to hadrons. Compare these values with those reported in Table 4.2.

6.19 Solutions

Solution to 6.1 Gravity and electrostatics have the same functional dependence on distance, so the ratio is the same at all distances. From Eq. (6.1), the ratio between the two forces is

$$R = \frac{F_G}{F_{EM}} = \frac{G_N\, m_e\, m_p\, 4\pi\, \epsilon_0}{Q_e^2}$$

$$= 6.674 \times 10^{-11}\ \mathrm{m^3 kg^{-1} s^{-2}} \times 9.109 \times 10^{-31}\ \mathrm{kg} \times 1.673 \times 10^{-27}\ \mathrm{kg} \times$$

$$\times\, 4 \times 3.14 \times 8.854 \times 10^{-12}\ \mathrm{F\, m^{-1}}/(1.602 \times 10^{-19}\ \mathrm{C})^2$$

Considering that a Farad is $F = \mathrm{s^2\, C^2\, m^{-2}\, kg^{-1}}$, the dimensional analysis gives

$$\mathrm{m^3 kg^{-1} s^{-2}\ kg\ kg\ s^2\ C^2\ m^{-2}\ kg^{-1}\ m^{-1}\ C^{-2}} = \text{dimension-less},$$

as expected. For the orders of magnitudes:

$$-11 - 31 - 27 - 12 + 2 \times 19 = -81 + 38 = -43$$

and for the numerical value

$$6.674 \times 9.109 \times 1.673 \times 4 \times 3.14 \times 8.854/2.57 = 4400$$

So the ratio between these two forces between an electron and a proton, at any distance, is 4.4×10^{-40}.

Solution to 6.2   The possible decay vertices are:

$$Z^0 e^+ e^-; \ Z^0 \mu^+ \mu^-; \ Z^0 \tau^+ \tau^-; \ Z^0 \nu_e \bar{\nu}_e; \ Z^0 \nu_\mu \bar{\nu}_\mu; Z^0 \nu_\tau \bar{\nu}_\tau;$$

$$Z^0 u\bar{u}; \ Z^0 d\bar{d}; \ Z^0 c\bar{c}; \ Z^0 s\bar{s}; Z^0 b\bar{b}; \quad Z^0 t\bar{t};$$

The last diagram corresponds to a legitimate vertex, but the phase space for this decay is zero because the mass of the Z^0 is lower than the mass of the *top* quark, so this decay does not occur. The Feynman diagrams corresponding to decays to quarks have to be multiplied by the number of colours. In total there are $6 + 15 = 21$ diagrams with non-zero phase space. In the approximation of this problem, all these processes are equally possible; therefore the branching fractions to $c\bar{c}$ is expected to be $3/21 = 14.3\%$, the branching fraction to hadrons is expected to be $5 \times 3/21 = 71.4\%$. The corresponding values in Table 4.2 are 12.03% and 69.91%, in quite a good agreement for this simplified calculation. This simple calculation also explains qualitatively how the width of the z^0 depends on the number of light neutrinos, as shown in Fig. 6.2: with more processes available and equally possible, the lifetime decreases, increasing the width.

Bibliography and Further Reading

G. Barr, R. Devenish, R. Walczak, T. Weidberg, *Particle Physics in the LHC era* (Oxford University Press, Oxford, 2016)

A. Bettini, *Introduction to Elementary Particle Physics* (Cambridge University Press, Cambridge, 2014)

S. Braibant et al., *Particles and Fundamental Interactions* (Springer, Dordrecht, 2012)

K. Gottfried, V. Weisskopf, *Concepts of Particle Physics*, vol. 1 (Oxford University Press, Oxford, 1984)

A. Kamal, *Particle Physics* (Springer, Berlin, Heidelberg, 2014)

B.R. Martin, *Nuclear and Particle Physics* (Wiley, Hoboken, 2009)

B.R. Martin, G. Shaw, *Particle Physics* (Wiley, Hoboken, 1992)

D. Perkins, *Introduction to High Energy Physics* (Addison Wesley, Boston, 1987)

B. Povh et al., *Particles and Nuclei: An Introduction to the Physical Concepts* (Springer, Berlin, Heidelberg, 2015)

M. Thomson, *Modern Particle Physics* (Cambridge University Press, Cambridge, 2013)

Chapter 7
Introduction to Nuclear Physics

7.1 Introduction

Atomic nuclei are extended objects (Fig. 7.1), like the neutrons and protons of which they are made: we can measure their size, while fundamental particles, like electrons and quarks, behave as point-like objects.

The force which binds together the nucleons within a nucleus is the strong nuclear interaction, the same that binds together quarks to form baryons and mesons. However, this force is now acting on a longer range and it can be pictured as being a residual force that is left after forming the bound states of quarks, which are the nucleons. Using an analogy, the nuclear strong force is similar to the electromagnetic attraction which binds together two neutral atoms to form a molecule. In our case, two colourless nucleons are bound together to form a nucleus. Of course, the interaction is completely different in the two cases.

The simplest stable bound state of two nucleons is the *deuteron*, which is made of a proton and a neutron, as shown in Fig. 7.2. The presence of the proton stabilises the neutron, making it energetically unfavourable to undergo a beta decay. The next simplest nuclide is the *triton*, or *tritium nucleus*, with two neutrons and one proton: ^{3}H. It is unstable and has a lifetime of 12.3 years, decaying β to ^{3}He, which is stable. The next simple nuclide is ^{4}He, which is extremely stable and is also known as α particle. At the opposite end, we have nuclides which are made of a very large number of nucleons: an example is Uranium, which contains 238 nucleons. Different models are used to describe, to various levels of approximation, these systems, which are made of a number of nucleons ranging between 2 and 240. Unlike what occurs for atoms, where the energy levels can be precisely calculated, each of the nuclear models is successful to explain some features, but not all.

© Springer Nature Switzerland AG 2018   147
S. D'Auria, *Introduction to Nuclear and Particle Physics*,
Undergraduate Lecture Notes in Physics,
https://doi.org/10.1007/978-3-319-93855-4_7

Fig. 7.1 Pictorial
representation of a nucleus,
made of neutrons and
protons. Clearly, it is only a
way to represent it and we
don't have to take it literally

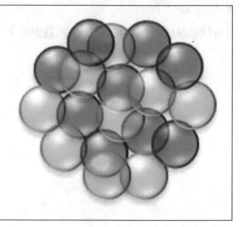

Fig. 7.2 Pictorial
representation of a deuteron,
bound state of (*pn*). It is a
stable nuclide, an isotope of
Hydrogen ^{2}H. Heavy water
has one deuterium atom
replacing hydrogen.
Deuterium is not an element
but is occasionally indicated
with D, and heavy water with
HDO

7.2 The *Q*-Value of a Reaction

The *Q-value* is defined as the difference between the kinetic energy of the final
system and the kinetic energy of the initial system:

$$Q = E_k(\text{final}) - E_k(\text{initial}) \tag{7.1}$$

In a generic reaction $A + B \rightarrow C + D$, we can write the energy conservation as:

$$m_A c^2 + K_A + m_B c^2 + K_B = m_C c^2 + K_C + m_D c^2 + K_D \tag{7.2}$$

where K_X indicates the kinetic energy of the particle X. Rearranging:

$$K_C + K_D - K_A - K_B = Q \qquad (7.3)$$

$$Q = m_A + m_B - (m_C + m_D) \qquad (7.4)$$

We therefore derive that the Q-value is the mass difference between the initial and final states (note the inverted order with respect to the kinetic energy). As a consequence, an elastic scattering $A + B \to A + B$ has $Q = 0$. An example of elastic scattering is Compton scattering. A decay can occur spontaneously only if $Q > 0$. Of course, this is not the only condition: also electric charge, total angular momentum, and other quantum numbers have to be conserved. Reactions with $Q > 0$ transform mass into kinetic energy; reactions with $Q < 0$ transform kinetic energy into mass. An example of the latter is the creation of pairs of top quarks in proton–proton collisions. The proton mass is ≈ 1 GeV/c^2, and the *top* mass is $m_t \approx 174$ GeV/$c^2 \gg m_p$. These reactions are "endothermic" or "endo-energetic", and they require a large kinetic energy to create a pair of particles with large mass.

We define the *threshold energy for a reaction to occur* as the minimum kinetic energy required to the initial particles, when the product particles have zero kinetic energy. If the particles in the initial state have a lower value of kinetic energy, the reaction cannot take place.

In this case:

$$K_A + K_B = (m_C + m_D) - (m_A + m_B) \qquad (7.5)$$

As an example, photons can interact with matter by creating an electron–positron pair, in the presence of a nucleus (to conserve momentum):

$$\gamma + (A, Z) \to e^+ + e^- + (A, Z) \qquad (7.6)$$

The photon has no mass, so all its energy is kinetic. The threshold of the reaction above is

$$K_\gamma = 2m_e = 2 \times 511 \text{ keV} \qquad (7.7)$$

In decays where $A \to C + D$ and $K_A = 0$, the requirement of $Q \geq 0$ translates into

$$m_A \geq m_C + m_D \,, \qquad (7.8)$$

that is, the mass of the parent must be larger than the combined mass of the daughters. The Q-value is connected mathematically to the factor ρ_f the phase space density of the final states, Eq. (6.9).

7.3   Atomic Nuclei Phenomenology

One of the properties to look for at first is the nuclear size and the nuclear
density. We'll address some stability issues and review one model of the atomic
nucleus, before showing some nuclear reaction of practical interest. The size of
the nuclei is experimentally measured by means of scattering experiments, which
probe the nucleus with particles. These can be either α or electrons accelerated to
medium or high energies up to a few MeV. The nuclear size, as it appears from
the electromagnetic interaction, is measured experimentally by electron–nucleus
scattering, which is sensitive to the electrical charge density. The nuclear radius
R turns out to be

$$R = 1.21A^{1/3}; \quad R \text{ in fm} \tag{7.9}$$

Assuming that nuclei have a spherical shape, their volume is $V = 4/3\pi R^3 = 7.42A$
(fm^3), and their density is 0.13 nucleons per fm^3. It is quite constant for $55 \leq A \leq$
209. To probe the nuclear density, as seen by the nuclear force, other particles can
be used as "projectiles": $\pi^\pm, \pi^0$ or neutrons. Various models of atomic nuclei can
be used, depending on the experiment and on the quantity being measured, like the
black disk in a model and the *optical model*. The mass of nuclei is expressed in
terms of the *atomic mass unit*, which is defined to be 1/12 of the mass of an atom of
^{12}C. We should note explicitly here that the mass of the electrons is included in this
unit. The a.m.u. is

$$u = 931.4940 \,\text{MeV/c}^2 = 1.6605389 \times 10^{-27} \,\text{kg.} \tag{7.10}$$

The mass of an atom, as determined experimentally, is less than the sum of the
masses of its components:

$$M(Z, N) < Z(m_p + m_e) + Nm_n \tag{7.11}$$

$$- B \equiv \Delta M(Z, N) \equiv M(Z, N) - Z(m_p + m_e) - Nm_n \tag{7.12}$$

We call ΔM the *mass deficit*; when multiplied by c^2, we obtain the total *binding
energy* B: if we want to separate completely all nucleons, we need to provide an
energy equal to the binding energy. The interesting quantity is the binding energy
per nucleon B/A. Its experimental value is plotted for stable nuclides in Fig. 7.3 as a
function of A. On average, the value of B/A is between 7 and 9 MeV per nucleon. It
increases for small nuclides, has a maximum for ^{56}Fe, then decreases. The nuclides
in the raising part of the curve can, under appropriate conditions, *fuse* with other
small nuclides and emit energy; in this case, we have a *nuclear fusion*. The nuclides
at high A can gain energy if they split into smaller nuclides; this is why, we have
nuclear fission.

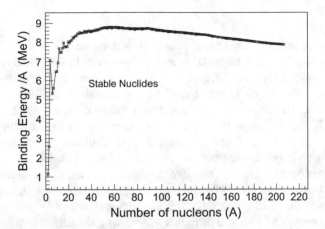

Fig. 7.3 Binding energy per nucleon as a function of the mass number $A = Z + N$ for stable or long-lived nuclides. Nuclides with low A and low binding energy per nucleon can undergo a *fusion* reaction, under appropriate conditions, to increase A and move towards higher binding energies. Conversely, nuclides with high A, when splitting into two nuclides with lower A, can increase the binding energy per nucleon in each of the daughter nuclides. Fission and fusion processes have $Q > 0$ for high and low A, respectively. It is worth noting that ^{4}He, or α particle, has a binding energy of 7 MeV per nucleon, making it a very stable nuclide (data from IAEA Nuclear Data Section)

7.4   The Liquid Drop Model

A spherical object with uniform density can be thought as resembling a liquid drop in the absence of gravity. We can use this model to describe nuclei and calculate their mass with the *semi-empirical* mass formula of Bethe–Weizsäcker[1] as the sum of six terms: the first is just the sum of the mass of components, while the other five represent the binding energy:

$$M(A, Z) = \sum m_i - a_1 f_1 + a_2 f_2 + a_3 f_3 + a_4 f_4 + f_5 \qquad (7.13)$$

$$= \left(Z(m_p + m_e) + N m_n\right) - a_V A + a_S A^{\frac{2}{3}} + \qquad (7.14)$$

$$+ a_C Z^2 A^{-\frac{1}{3}} + a_A (A - 2Z)^2 A^{-1} + f_5 \qquad (7.15)$$

where the numerical values in MeV are

$$a_V = 15.76; \quad a_S = 17.81; \quad a_C = 0.711; \quad a_A = 23.7 \qquad (7.16)$$

---

[1] After Hans Bethe (1906–2005) and Carl Friedrich von Weizsäcker, (1912–2007). Both were from Germany, Bethe moved to Manchester, Bristol (UK) and later to Cornell, USA, because of racial laws.

The single terms are

- a mass term from the constituents: $Z(m_p + m_e) + N m_n$;
- a volume term, which is negative and proportional to A; it accounts for the average binding energy between a nucleon and its immediate neighbours.
- a surface term, which is positive and proportional to $A^{2/3}$, accounts for the nucleons at the surface, which have fewer neighbours and therefore are less bound. If the radius is proportional to $A^{1/3}$, the surface of the sphere is proportional to $A^{2/3}$. Together with the next Coulomb interaction term, the surface term limits the maximum value of A for stable nuclei; this surface term gives the name to the model, it is like the surface tension of a liquid drop.
- the Coulomb term accounts for the mutual repulsion of protons; it is positive, as it decreases the binding energy, and is proportional to $Z^2 A^{-1/3}$. This term is the potential energy of a sphere of radius $r_0 A^{1/3}$ with a uniform charge $+Ze$. The theoretical value for $a_c = 0.714\,\mathrm{MeV}$ matches remarkably well the experimental value in Eq. (7.16).
- the asymmetry term: nuclei tend to be stable when $Z \approx N$ (Fig. 7.4) so the term is

$$f_4 = a_A \frac{(Z - A/2)^2}{A}$$

Fig. 7.4 Segrè chart of the stable nuclides: they tend to have $N \approx Z$ for low $Z \leq 20$, while heavier elements tend to have a slight neutron excess. This plot, together with Figs. 7.5 and 7.6, is the experimental basis of the asymmetry term in the semi-empirical mass formula (data from IAEA Nuclear Data Section)

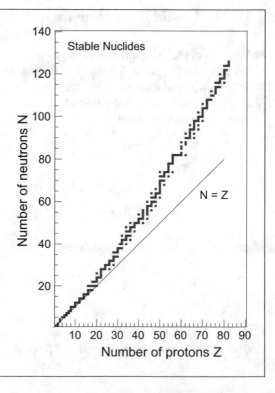

- the *pairing term*: inside a nucleus, nucleons tend to form pairs (nn) and (pp) with opposite spin. Thus, nuclides with Z even and N even have more binding energy than nuclides where there is an odd number of either or both nucleons. Only four nuclides with both Z and N odd are stable, while 167 stable nuclides have both Z and N even. The term in the formula will be positive for A, N odd, zero for even–odd or odd–even, and negative for both A, N odd. This term has a purely empirical value of

$$f_5 = \pm 12 A^{-1/2} \text{MeV/c}^2 \text{ (+for odd-odd)}; \quad f_5(\text{even,odd}) = 0 \qquad (7.17)$$

7.5 Beta Decays

The pairing term in Eq. (7.17) (Figs. 7.5 and 7.6) explains why some nuclides undergo β decays: the total energy is lower, or the binding energy is higher, if a neutron is replaced with a proton: in this case, we have a β^--emitting transition; when a lower-energy state is reached by replacing a proton with a neutron, the nuclide decays emitting a β^+. The decay

$$p \rightarrow n + e^+ + \nu_e \qquad (7.18)$$

in which

$$u \rightarrow d + e^+ + \nu_e \qquad (7.19)$$

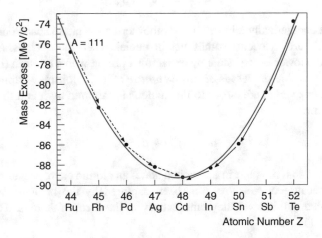

Fig. 7.5 Mass excess, which is minus the mass deficit, for the isobars with $A = 111$. These are all even–odd or odd–even nuclides, for which the pairing term is zero. The circles are the experimental values, the parabola is the calculation of the semi-empirical mass formula (SEMF). The arrows indicate beta decays: arrows towards the left hand side represent β^+ decays ($Z' \rightarrow Z - 1$), arrows towards the right-hand side represent β^- decays ($Z' \rightarrow Z + 1$). Data from IAEA Nuclear Data Section, plot modified after B.R. Martin, *Nuclear and Particle Physics*

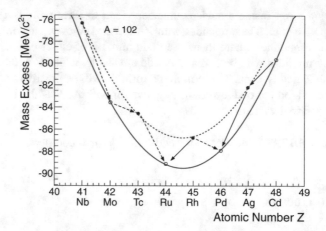

Fig. 7.6 Mass excess (or minus mass deficit) for the isobars $A = 102$. The experimental values now lay on two parabolas, and therefore the *pairing term* was introduced. Open circles represent (Z, N) even–even nuclides, filled circles represent (Z, N) odd–odd nuclides. Both ^{102}Ru and ^{102}Pd are stable nuclides. β^- transitions are indicated with a dashed arrow, and β^+ transitions with a full line. ^{102}Rh can undergo both transitions. Out of all the stable nuclides, only four have odd numbers of both protons and neutrons

is perfectly "legal" from the point of view of conservation laws. It has the same matrix element $|\mathcal{M}_{if}|^2$ as the "standard" beta decay:

$$d \rightarrow u + e^- + \overline{\nu}_e \tag{7.20}$$

but it is not energetically allowed, or in other terms the phase space density of the final state is zero. It occurs sometimes in nuclei, when the nucleus, as a whole, would reach a lower-energy state by replacing a proton with a neutron (Fig. 7.7).

Another possible process is *electron capture (Fig. 7.8)*: it may happen that one of the electrons which are closer to the nucleus is captured by it and the following reaction occurs:

$$p^+ + e^- \rightarrow n + \nu_e \,. \tag{7.21}$$

In this case, the atom is left in an excited state: an atomic energy level has become available, because one electron has been captured by the nucleus. When other electrons occupy this level they emit energy in the form of X-rays. The reaction at quark level is

$$u + e^- \rightarrow d + \nu_e \tag{7.22}$$

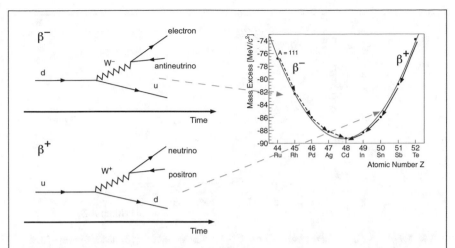

Fig. 7.7 Feynman diagrams at the quark level for the β^- (upper left) and β^+ decays (lower left). The two decays have almost identical Feynman diagrams, and therefore they have the same value for the matrix element $\mathcal{M}_{if}$. To determine whether the nuclide is stable or decays β^- or β^+, we have to calculate the energy variation related to this transition which is shown in the plot on the right-hand side. Isolated neutrons undergo β^- decays, because they have a mass which is slightly larger than the sum of the proton, electron and neutrino mass, i.e. the Q-value of the decay reaction is > 0. Protons are stable, because the β^+ decay is not energetically allowed, having a Q-value < 0. Inside a nucleus, the transition rate depends on the Q-value of the reaction at nuclear level, which depends on the pairing factor in the SEMF

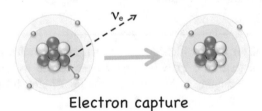

Electron capture

Fig. 7.8 Schematic representation of electron capture in the reaction $^7\text{Be} + e^- \rightarrow {}^7\text{Li} + \nu_e$

7.6 Double Beta Decay

Double beta decay can occur in even–even nuclei when the decay $(A, Z) \rightarrow (A, Z + 1)$ is energetically forbidden, but $(A, Z) \rightarrow (A, Z + 2)$ is energetically allowed, as shown in Fig. 7.9. It is an extremely rare decay because two almost independent transitions have to occur at the same time. It was observed for the first time in 1987, in the decay

$$^{82}_{34}\text{Se} \rightarrow {}^{82}_{36}\text{Kr} + 2e^- + 2\bar{\nu}_e \tag{7.23}$$

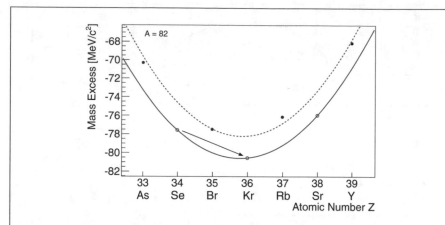

Fig. 7.9 Mass excess for the isobars with $A = 82$. The values for $^{82}_{34}$Se and $^{82}_{35}$Br are -77.594 and -77.496 MeV/c^2, respectively, so the decay $^{82}_{34}$Se $\rightarrow ^{82}_{35}$ Br $+ \beta^- + \bar{\nu}_e$ is not allowed

and it has been observed in the decay of ten other nuclides since then. Double electron capture has not been observed directly, but there is indirect evidence of it in geochemical samples. It may be possible in

$$^{102}\text{Pd} + 2e^- \rightarrow {}^{102}\text{Ru} + 2\,\nu_e \tag{7.24}$$

This transition is energetically allowed, as shown in Fig. 7.6, while ^{102}Pd cannot decay to ^{102}Rh.

The category of fermions which are their own anti-particles is named after Ettore Majorana (b. 1906, ?), who first advanced the hypothesis of their existence. He went missing in March 1938, during or after his boat trip from Palermo to Naples, Italy.

Neutrinoless double beta decay has never been observed: it may be possible if neutrinos are *Majorana particles*. This category of particles refers to massive fermions which are their own anti-particles. In the case of neutrinoless double beta decay, the electron energy spectrum would be a characteristic single-energy line. The neutrinoless double beta decay, if exists, would violate the lepton number conservation (Eq. (6.46)) in the sector of charged leptons.

7.7 Other Models

Other models for the nucleus are the *Fermi gas model*, the *shell model* and the *collective model*.

In the Fermi gas model, nucleons are considered as non-interacting particles.

Their mutual interaction is replaced by the resultant of their attraction, which forms a potential well and keeps nucleons inside a sphere of radius R_0, as shown in Fig. 7.10. These non-interacting nucleons are subject to the Fermi statistics, which gives the name to the model, and, therefore, to the Pauli exclusion principle. Let's start considering neutrons only: any quantum state, or energy level, in the well can be occupied by two neutrons, with opposite spin. The same is valid for protons. Now, while the wells of the two types of nucleons are not required to have the same depth, it is assumed that they have the same radius. In addition, the proton potential has a different shape, to account for the electrostatic repulsion outside the well. Nucleons occupy all available energy states, up to a level E_F, which is called "Fermi energy level".

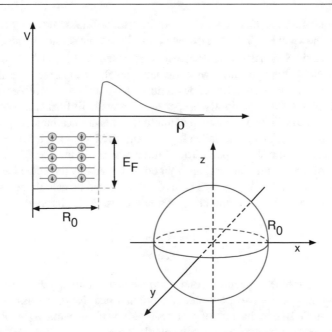

Fig. 7.10 A representation of the potential well for protons in the Fermi gas model. Protons are considered as non-interacting and are confined inside a sphere of radius R_0, the potential well is one-dimensional along the polar coordinate ρ; all available states are occupied up to the energy $E_F^{(p)}$. The Coulomb potential $1/\rho$ is sketched for $\rho > R_0$

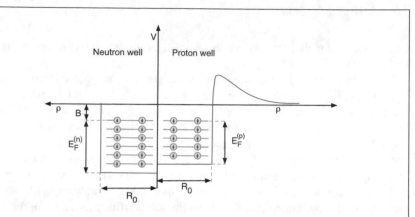

Fig. 7.11 The potential well for neutrons (left) and protons (right) in the Fermi gas model: all states are occupied up to the same energy level, but the depth of the two potential wells may be different

In a stable nucleus, this energy is required to be the same for protons and neutrons, as shown in Fig. 7.11, where the two wells are shown side by side, with the same vertical axis. When the two Fermi levels are not equal $E_F^{(n)} \neq E_F^{(p)}$, the nuclide will reach a stable configuration, with the same level, by undergoing a $\beta^{\pm}$ decay. The fact that at high A stable nuclides tend to have slightly more neutrons than protons is explained with a slightly deeper neutron well. To further advance in this model, we need to calculate how many quantum states are available, separately for protons and neutrons, in a potential well. In Quantum Mechanics, a state occupies a fundamental cell in the phase space. The volume of this cell is $h^3 = (2\pi\hbar)^3$. This is three orders of magnitudes larger than what could be naively expected based on the uncertainty relations $\Delta x\,\Delta p_x \geq \hbar/2$ (Eq. (6.2)) but it can be calculated using basic quantum mechanics. So, the number n_q of quantum states is simply

$$n_q = 2\frac{(4/3\pi\,R_0^2)(4/3\pi\,p_F^3)}{(2\pi\hbar)^3}\,, \tag{7.25}$$

where p_F is the momentum corresponding to the state of energy E_F and the factor of two accounts for the two possible spin orientations. Now, we assume that all states are occupied, so in the case of protons we know already that $n_q = Z$, and that $R_0 = r_0 A^{1/3} = 1.21\ \mathrm{fm} A^{1/3}$, so we can calculate p_F:

$$Z = \frac{32}{9}\frac{\pi^2 r_0^3 A p_F^3}{8\pi^3\hbar^3} = \frac{4\,A}{9\,\pi}\left(\frac{r_0 p_F}{\hbar}\right)^3$$

$$p_F = \frac{\hbar}{r_0} \sqrt[3]{\frac{9}{4}\pi \frac{Z}{A}} \approx \frac{\hbar}{r_o} \sqrt[3]{\frac{9}{8}\pi} = \frac{1.52 \times 0.658}{1.21} \quad \text{eV/(m/s)} \tag{7.26}$$

We have used the approximation: for stable nuclides $Z/A \approx 1/2$. Now, momenta are normally measured in MeV/c, where c is the speed of light, so we need to multiply the result above by an a-dimensional 3×10^8 to obtain

$$p_F \approx \frac{1.52 \times 0.658 \times 299{,}792{,}458}{1.21} \approx 250\,\text{MeV}/c \tag{7.27}$$

The proton mass is $m_p = 938.2720\,\text{MeV}/c^2$ and therefore

$$\beta\gamma = \frac{p}{mc} = \frac{250}{928} = 0.26 < 1$$

protons inside nuclei can have quite a high momentum, but there is no need to use relativity. The Fermi energy can be calculated as:

$$E_F = \frac{p_F^2}{2m_p} = \frac{250}{2 \times 938} = 33\,\text{MeV} \tag{7.28}$$

The difference between E_F and the top of the potential well is the binding energy B, which for nuclides with $A > 25$ is about 8 MeV per nucleon, as shown in Fig. 7.3, therefore the depth of the potential well is

$$V_0 = E_F + B = 33 + 8 = 41\,\text{MeV}. \tag{7.29}$$

The shell model was proposed separately by Maria Göppert-Mayer (1906–1972), who was in Chicago, and Hans Jensen (1907–1973), who was in Heidelberg, in 1950. They were awarded the Nobel prize in 1963.

This model can explain quite well also the asymmetry term of the mass formula. However, there are features of nuclides that require a more refined model. It was observed that the abundance of elements and their stability is maximum for some particular values of Z and $(A - Z)$: 2, 8, 20, 28, 50, 82, and 126. These numbers have been called *magic numbers* before theory could explain them. This effect is similar to what happens with atoms, where completely filled electronic shells correspond to chemically stable and inactive elements, the six noble gases. Without entering into details, qualitatively the shell model adds a *spin–orbit* interaction to the square potential well of the Fermi gas model. The strong interaction between nucleons becomes now spin-dependent, along the same lines as the electromagnetic spin–orbit interaction, which slightly modifies the atomic energy levels. By adding a spin–orbit term to the potential,

$$V = V_{\text{well}}(r) + V_{\text{so}}(r)\,\vec{j} \cdot \vec{L}\,, \tag{7.30}$$

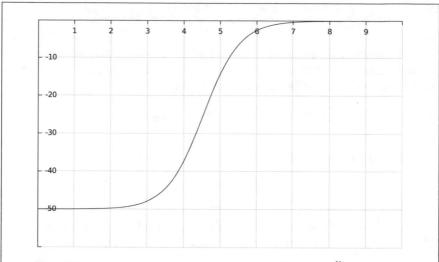

Fig. 7.12   The Woods–Saxon potential well for neutrons: $V(r) = -\dfrac{V_0}{(1+e^{\frac{r-R}{a}})}$

where $\vec{j}$ indicates the spin and $\vec{L}$ the orbital angular momentum, the model can explain all seven magic numbers.

Further improvements modify the shape of the potential well, as shown in Fig. 7.12. Some nuclides are extremely stable as they have both proton and neutron shell completed, like noble gases. Te α particle is one example, at the other end of the spectrum is the most abundant isotope of lead $^{208}_{82}\text{Pb}_{126}$.

7.8   Alpha Decays

At this point, we can try to understand the mechanism of alpha decays. Its quantitative explanation requires quantum mechanics. The *tunnelling* effect has no equivalent in classical physics: a potential barrier of finite width can be crossed by a particle even if its energy is lower than the potential barrier. The probability for this process to occur depends on the width of the barrier at the energy level of the particle. It is outside the scope of this introductory book to calculate the decay rate with Gamov's tunnelling, see, e.g. Martin (2009). Remaining on the qualitative description, we first notice from Fig. 7.3 that emitting a nucleon is energetically allowed ($Q > 0$) only for heavy nuclides, with $Z \geq 105$. The α-emitting nuclide with lowest A is $^{105}_{52}\text{Te}$. The next question is why we observe α emission and not single neutron or single proton emission. Neutron emission by *evaporation* from a nucleus, many contributions fromes from a further analogy to the liquid drop, is a process that occurs normally, e.g. when fission fragments are produced. However, it is a very rapid process, which immediately follows the formation of new nuclei. Apart from this, there is an easier way for the nucleus to reach a lower-energy state, and this is β decay, which transforms neutrons into protons and vice versa. Very often, α and β emissions are competing processes, or different decay branches of the same nuclide. Alpha decay lifetimes range from 8×10^{-7} s to 6×10^{26} s. An alpha particle is a "doubly magic" nuclide ($Z = 2$, $N = 2$), with a binding energy of 7 MeV per nucleon, which is close to the maximum. In the *collective model* of heavy nuclei, we can consider them as formed by a tightly bound core and an outer layer, which may be modeled as a liquid drop, not necessarily spherical. For large nuclei, a part of this outer shell can be modeled as containing an α particle. In some very recent models also, some nuclear states of nuclides like ^{8}Be and ^{12}C are considered as α particle *condensates* (see, e.g. Yamada (2012)) of, respectively, two and three alpha particles, each α being a spin zero boson.

The alpha particle inside an unstable nucleus can be modeled as being in a potential well (Fig. 7.13); as it is electrically charged, the potential well is like the one of protons (Fig. 7.10), with a Coulomb barrier outside the well. We suppose that the α is formed among the most energetic nucleons with a certain probability. If the energy level of this alpha particle inside the nucleus is below the zero level, the α decay will not occur. If, however, the energy level is above zero, there is possibility of tunnelling through the Coulomb barrier. Alpha particles from a given nuclide are emitted with a single energy, which is often a marker to identify the nuclide that has emitted it. From the experimental point of view, the lifetime of pure alpha emitters is linked to the energy of the alpha particle by the Geiger–Nuttall relation:

$$\log \tau = a + b \frac{Z}{\sqrt{E_\alpha}} . \tag{7.31}$$

Fig. 7.13 Some nuclides can be modeled as having α particles as one of the components. If the energy level of this alpha particle is larger than zero, it can *tunnel* through the Coulomb barrier and be emitted

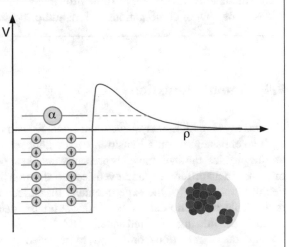

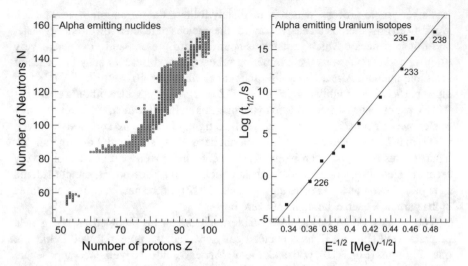

Fig. 7.14 Left: Segrè chart of the α-emitting nuclides: they all have $A \geq 105$. Right: the Geiger–Nuttall plot for isotopes of uranium, showing a linear relation between Log_{10} of the lifetime in seconds and the inverse of the square root of the kinetic energy. Nuclides with short lifetime emit more energetic α particles. The lifetime spans 20 orders of magnitude, while the α kinetic energy varies within a few MeV (data from IAEA Nuclear Data Section)

This relation can be derived with Gamow's tunnelling model (see, e.g. Martin (2009)) and is in good agreement with data, as shown in Fig. 7.14.

George Gamow (Russia 1904, USA 1968) taught in Leningrad (now St. Petersburg), then moved to Europe end then to the USA, where he taught in various universities. He is known for many contributions from α decays to cosmology, DNA combinations and for authoring many popular science books.

7.9   Gamma Emission

Gamma rays are high-energy quanta of light. We can use the shell model to explain such an emission, when a transition occurs between nuclear energy levels. Unlike atomic physics, the shell model is presently unable to predict the exact energy levels, or rather their difference, and even less so the transition amplitudes. However, a detailed "mapping" of the experimentally measured energy levels for each nuclide is available in public databases. Both α and $\beta^{\pm}$ decays in most cases leave the daughter nuclide in an excited state.

Subsequent transitions, or decays, to intermediate states and to the ground state are marked by emission of one or more gamma rays. This is very similar to what

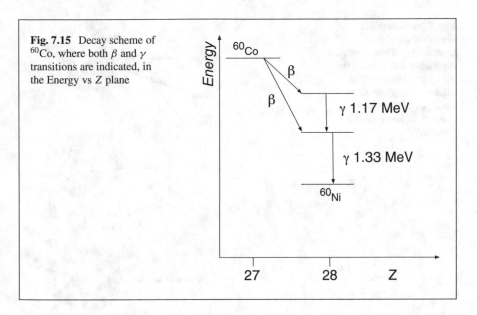

Fig. 7.15 Decay scheme of ^{60}Co, where both β and γ transitions are indicated, in the Energy vs Z plane

happens in atomic physics, where light emission occurs upon transition from excited electron states. Similarly to atomic physics, nuclear energy levels are indicated as horizontal segments in plots where the vertical axis represents the energy and the horizontal axis represents A or Z, as shown in Fig. 7.15. State transitions which emit γ's are indicated as vertical arrows, they do not change A or Z, while α and β emissions are indicated as arrows with a slope in the $(E - Z)$ plane. Also, the spin associated with the level can be indicated, together with the energy level and the lifetime or branching fraction of the transition.

7.10   Passage of Radiation Through Matter: Neutrons

Charged particles lose their kinetic energy by ionising the atoms: they interact with the electrons. Heavily ionising particles are either highly electrically charged, like alphas and fission fragments, or slow, or both. They lose all their energy and are stopped by a thin amount of solid material. High-energy, penetrating particles like *muons* ($\mu^{\pm}$) lose part of their energy with the same ionisation mechanism and are deviated in their trajectory by the electric field of the nuclei. The Bethe–Bloch equation (5.38) allows us to calculate the energy loss for charged particles. So far, we have not considered neutral particles. The only two types of neutral particles which live long enough to interact with matter are neutrinos and neutrons. Neutrons are particularly important, because they can provoke nuclear fission. Being neutral, they do not interact with the electrons of the atoms. They are baryons, so they have a residual strong force and they interact with the nuclei (Fig. 7.16).

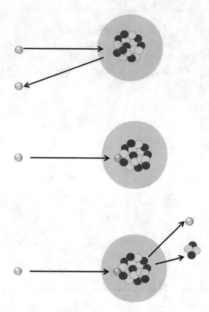

Fig. 7.16 Schematic visualisation of the interactions of neutrons with the matter. All interactions occur with nuclei: elastic scattering (top), neutron absorption (middle) and generic nuclear reactions (bottom)

There are three possible reactions:

- elastic scattering $n + A \rightarrow n + A$
- capture $n + A \rightarrow (A + 1)$
- other nuclear reactions: $n + A \rightarrow B + C$

Elastic Scattering

In the elastic scattering process, neutrons lose their energy by transmitting part of it to the nuclei. Both parts remain unchanged by the scattering. For kinetic energies lower than about 1 MeV, this is the only possible scattering process, and it can be treated mathematically just as a classical billiard balls scattering, with balls of different mass. Neglecting the mass difference between protons and neutrons, and the binding energy, A represents the atomic mass. We have maximum energy transfer when the collision occurs in one dimension: the neutron bounces back, and the nucleus recoils along the initial direction of the neutron. The heavier the nucleus, the lower fraction of neutron energy is transmitted to it. The minimum energy transfer is zero for extremely peripheral collisions. After one scattering event, the neutron has a kinetic energy E which is a fraction of the initial energy E_0

$$\left(\frac{A-1}{A+1}\right)^2 \leq \frac{E}{E_0} \leq 1 \,. \tag{7.32}$$

All energy losses within this range are equally probable. In case of scattering on hydrogen, $A = 1$ after n scattering processes the neutron average kinetic energy is:

$$< E_n >= \frac{E_0}{2^n} .$$ (7.33)

Neutron Capture

In the capture process, the neutron remains bound to the nucleus, which now has increased by one its mass number $A + 1$. The atomic number is unchanged, unless, of course, there is a subsequent beta decay. A very important neutron capture process is

$$n + {}^{113}\text{Cd} \rightarrow {}^{114}\text{Cd}^* \rightarrow {}^{114}\text{Cd} + \gamma .$$ (7.34)

This is used in nuclear reactors to control the flux of neutrons in the core. Another example of capture, which occurs in nuclear reactors, produces radiocarbon:

$$n + {}^{13}\text{C} \rightarrow {}^{14}\text{C} + \gamma .$$ (7.35)

Other Reactions

Other reactions include the charge exchange scattering:

$$n + {}^{14}\text{N} \rightarrow p + {}^{14}\text{C}$$ (7.36)

which is the main reaction to produce radiocarbon in the atmosphere. In this case, neutrons originate from cosmic rays and have high energy. The same reaction can occur in a nuclear plant and in this case it has a cross section of 1.8 b, for thermal neutrons.

In a generic nuclear reaction, anything can happen, provided that all conservation laws are respected. Another important reaction is

$$n + {}^{10}\text{B} \rightarrow \alpha + {}^{7}\text{Li}^* + \gamma (0.48 \text{ MeV}) \quad (\sigma = 3480 \text{ barn})$$ (7.37)

The cross section for the above process, for very low-energy neutrons ($E_k \approx 0.025$ eV), is 3480 barn. Also, this reaction is used to remove neutrons from the core of a nuclear plant, to control the reaction.

The neutron-induced fission was discovered by Lise Meitner and Otto Hahn, in 1938–1939. Meitner was the first woman to become a full professor in physics in Germany; she had to leave in 1938 to Sweden. Hahn was awarded the Nobel prize for chemistry in 1944; the artificial element *meitnerium* $_{109}$Mt is named after L. Meitner. The most stable isotope $^{278}_{109}$Mt has a lifetime of about 10 s.

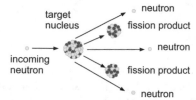

Fig. 7.17 A schematic representation of the neutron-induced nuclear fission. Notice that one or more neutrons are produced, which is essential for a self-sustained nuclear reaction

7.11   Spontaneous and Induced Fission

We have a spontaneous fission when a nucleus breaks into two daughter nuclei of approximately equal mass without any external action. As shown in Fig. 7.3, it occurs only for some nuclides with $A > 100$, which reach a higher binding energy per nucleon if they move to a lower A.

$$^{238}U \rightarrow {}^{145}La + {}^{90}Br + 3n; \quad \mathcal{BR} \approx 10^{-7} \tag{7.38}$$

It is a fairly rare process, which "competes" in terms of probability, or decay branching ratio, with α-emission. The associated production of one or more neutrons is very important, because they can in turn start a neutron-induced fission, which is at the basis of nuclear reactors. The neutron-induced fission is a form of nuclear reaction, whose effect is breaking a nucleus into two or more daughter nuclei of about the same mass; an example is

$$^{235}U + n \rightarrow {}^{92}Kr + {}^{141}Ba + 2n \tag{7.39}$$

Many odd-A nuclides are *fissile*, i.e. they have a large probability to undergo fission when a low-energy neutron interacts with the nucleus (Fig. 7.17). This is the case for ^{235}U, ^{239}Pu and ^{241}Pu.

Certain even-Z/even-N nuclides, like ^{232}Th, ^{238}U and ^{240}Pu, require energetic neutrons to undergo a fission process.

Enrico Fermi (1901–1954) from Italy, was where he received the prize in 1934 for his work on neutron-induced radioactivity. From Stockholm he moved to the USA, where he taught at Columbia University. He designed and operated the first nuclear reactor, which reached criticality in Chicago on December 2nd 1942. He also taught at the Universities of Rome and Chicago. More than 25 physics effects and models are named after him (photo: courtesy of Argonne National Laboratory)

7.12 Applications: Fission-Based Nuclear Reactors

The principle of fission reactors is based upon neutron-induced fission and an accurate neutron balance. The most stable uranium isotopes have $A = 238$ and $A = 235$. The former, ^{238}U, is more abundant in nature, but the probability that upon meeting a slow neutron it breaks into fission fragments is very small. On the contrary, ^{235}U only makes 0.7% of natural uranium, but the *cross section* for the reaction:

$$^{235}U + n \rightarrow \text{ fission fragments} + N_n \, n \qquad (7.40)$$

is a factor 10^8 larger. The cross section depends on the neutron kinetic energy, and it is larger for low-energy neutrons. When the energy of the neutrons is of the same order of magnitude of the average thermal energy kT, where k is the Boltzmann's constant and T the temperature in K, these neutrons are called *thermal* ($kT \approx 0.025$ eV). The fission fragments are not unique, as many different fission reactions can take place. They are often radioactive and most of them decay with medium-slow lifetimes. On average, the fission fragments carry 180 MeV per fission, the neutrons carry about 2.5% of the energy, and an additional 13% is obtained by a later-stage decay of radioactive nuclides. This is a significant fraction of the heat created with the fission. The kinetic energy is transformed into heat:

The energy loss for charged particles is calculated with the Bethe–Bloch formula, Eq. (5.38), while neutrons lose their kinetic energy by elastically scattering, as in Eq. (7.32), or by reacting with nuclei and producing heavy charged particles. A cooling system, typically water-operated, generates steam for a turbine.

Boltzmann's constant links the average energy of a gas to its temperature: $k = 8.61733 \times 10^{-5} \text{eV K}^{-1}$. It is named after Ludwig Boltzmann 1844–1906, from Vienna.

The key point to the nuclear *pile* is that the fission reactions generate neutrons, which in turn can induce fission to other ^{235}U nuclei. The number N_n of neutrons generated in each fission process, Eq. (7.40), is very important to establish a *chain reaction*. These neutrons are not thermal, and they carry quite a substantial amount of energy. The cross section for a fission reaction is much larger for thermal neutron. In order to thermalise them, a *moderator* has to be used. A moderator is some inert material which absorbs energy from fast neutrons, without absorbing them. The main process is the elastic scattering and therefore hydrogen, with $A = 1$, is the best possible moderator, from Eq. (7.32). Water and graphite are used as moderators as well as heavy water, which contains deuterium. The chemical symbol D is used for deuterium, and heavy water is also indicated with HDO. With reference to Fig. 7.18, we define the *criticality of a chain reaction* as:

$$k = \frac{\text{number of useful neutrons produced at stage}(n)}{\text{number of useful neutrons produced at stage}(n-1)} \qquad (7.41)$$

Neutrons are *useful* in a reactor when they don't escape, they are not absorbed by other material which may be present, but, after some multiple elastic scattering, they induce a fission process. We have a self-sustained and steady reaction when $k = 1$. At this point, the reactor has reached criticality. The reaction continues as long as there is nuclear fuel (^{235}U). We have an explosive reaction if $k > 1$: this is the principle of the nuclear bomb. The reaction will not continue if $k < 1$; in this case, the reactor is said to be *sub-critical*. In order to operate a reactor, we need to have a way to regulate and control the number of useful neutrons. This is done thanks to the properties of cadmium and boron, which we have seen in the reaction (7.34). There are three stable isotopes of Cd, and ^{114}Cd has a neutron capture cross section of about 2500 barn for thermal neutrons. For this reason, when cadmium rods are inserted in the nuclear fuel, they can remove neutrons and control the criticality of the nuclear reaction. Also, boron is used for the same purpose.

An example of fission reaction is

$$^{235}\text{U} + n \rightarrow\, ^{144}_{56}\text{Ba} +^{90}_{36}\text{Kr} + 2\,n \qquad (7.42)$$

In this case, the Q-value is 180 MeV. This is a small amount of energy from a macroscopic point of view, but we need to consider the huge number of atoms in a

Fig. 7.18 Representation of
a chain of fission reactions,
initiated by a neutron.
Different stages represent
consecutive time intervals

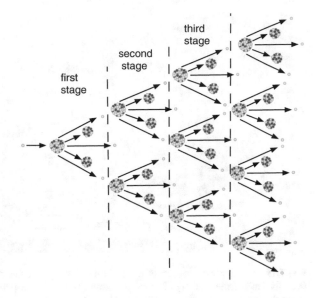

mole. In case of natural Uranium, a mole is 238 g; it contains $N_A = 6.02 \times 10^{23}$
atoms. Of these, 0.7% are of the fissile isotope ^{235}U. The total amount of energy we
can obtain from them is

$$W_T = 180 \times 10^6 \text{eV} \times 0.007 \times 6.0 \times 10^{23} \times 1.6 \times 10^{-19} \text{ J/eV} = 1.2 \times 10^{11} \text{ J}$$

Considering that $1 \text{ J} = 2.78 \times 10^{-7}$ kWh, the energy above can be expressed in
a more practical unit as 3.36×10^4 kWh. This energy is mostly in the form of
heat, although some of it, about 2%, escapes as neutrinos. The efficiency to convert
heat to electric energy in nuclear plants is about 30%, obtaining 11,200 kWh. A
household in Europe uses, on average, 3600 kWh of electricity a year. This makes
a mole of natural uranium (238 g) just about sufficient for the electrical needs of
three households in Europe for a year; it is equivalent to the energy of 2200 kg of
liquefied natural gas. The total yearly production of uranium is about 6×10^4 kg at
present. It is produced as "yellow cake" oxide, or U_3O_8.

Present nuclear reactors have somehow a "standard" size to produce a maximum
of about 1 GW of electrical power. More than one reactor unit can be clustered in
the same nuclear power plant, to share services, cooling and optimise the transport
of fuel and spent fuel. At present, 450 reactor units are operating in the world
(source: IAEA, 2016), producing a maximum electrical power of 390 GW. Different
technologies are used for power plant reactors. They can be characterised by the
nature of the moderator, by the coolant, by the type of fuel and by the type of nuclear
reaction that occurs.

- *Pressurised Water Reactors (PWR)* They are moderated and cooled by normal
  water, a schematic is shown in Fig. 7.19. The pressurised vessel avoids that

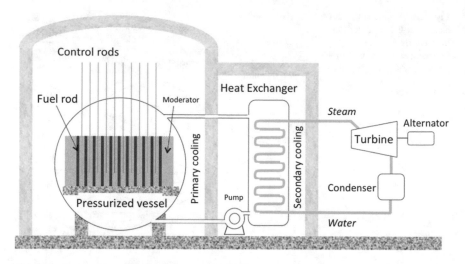

Fig. 7.19 A simplified diagram of a Uranium-based nuclear fission reactor, of the type PWR. (Modified after Lilley (2001).) For safety reasons, the coolant which circulates in the reactor core is not the same fluid which powers the turbine. The moderator is indicated as a separate element, but in Pressurised Water reactors the same pressurised water for primary cooling also acts as a moderator. CANDU reactors have a similar scheme, but the coolant in the primary circuit is heavy water (HDO)

coolant water boils inside the reactor core. For this reason, there are two cooling circuits, a primary and secondary.

- *Boiling Water Reactors (BWR)* only have one cooling circuit, otherwise they are very similar. They both use enriched uranium as fuel. This means that in the fuel rods the concentration of fissile isotope ^{235}U is increased with respect to the natural 0.7%. The fuel chemical compound is UO_2.
- *Heavy Water Reactors (HWR)* they are moderated and cooled by heavy water (HDO), where one atom of hydrogen is replaced by an atom of deuterium (^2H). The CANDU (CANadian Deuterium Uranium) reactors use natural uranium UO_2, and they can operate with a percentage of thorium. The schematics is very similar to the BWR.
- *Graphite Moderated Reactors* use the same moderator as the first reactor in Chicago. They can use either gas as a coolant (CO_2, He or N_2) or water, as in the case of the Chernobyl reactor.
- *Breeder Reactors* don't require a moderator and produce more fissile material than they consume. They are cooled with a molten salt, or with liquid sodium, as in case of the Superphenix reactor, which was operated in France. Their fuel is a mixture of ^{239}Pu which is fissile, and ^{235}U, which is *fertile*. The main reaction in a Pu-U breeder reactor are

$$^{239}Pu \rightarrow 3n + F_1 + F_2, \qquad (7.43)$$

$$n +_{92}^{238}U \rightarrow _{92}^{239}U \xrightarrow{\beta} _{93}^{239}Np \xrightarrow{\beta} _{94}^{239}Pu \qquad (7.44)$$

where F_1 and F_2 are fission fragments.

New types of reactors are being developed as "Generation IV", where new cooling schemes and type of fuel are being tested with the aim of improving efficiency and safety.

Although nuclear fission reactors don't produce directly any greenhouse gas, they have many other inconveniences. The main problem is production of nuclear waste, which is radioactive and has a decay time of million years. The other non-negligible problem is proliferation of nuclear armament: fission nuclear reactors can be used to produce nuclides which are most suitable for bombs. Other aspects are more common to any other human activity which involves large quantities of energy: operation safety, and protection from extreme natural events like earthquakes and tsunamis, and human mistakes. An additional factor is the magnitude and duration of the consequences that a possible accident would produce, which is far beyond the worst accident which could occur in a non-nuclear plant with a similar energy content. The consequences of the major nuclear plant accidents extended beyond the border of single countries: the release of radioactivity reached locations which were thousands of kilometres away from the plant, and entire cities had to be evacuated for decades.

The two most severe accidents to nuclear plants were the Chernobyl accident, which occurred on 26 Aprseventeen small naturalil 1986 near Pripyat, Ukraine, and the Fukushima Daiichi accident, occurred on 11 March 2011 in Japan. The first was due to a series of human mistakes, the second to a tsunami following an earthquake of magnitude 9.

Nuclear waste can be divided into three categories: fission products, *actinides* and activated material. Fission products contribute for a weight that is about the same, or slightly lower, than the amount of initial fissile material. If fuel is enriched to 3.5% of fissile uranium, from 0.7% of the natural uranium, we expect about 3% of fission products. Some of them have a relatively short lifetime, like ^{131}I, while others have a lifetime comparable to the human life. This is the case of ^{137}Cs and ^{90}Sr. However, some have a very large lifetime: ^{135}Cs and ^{129}I decay in 2.3 and 15 million years, respectively. The large flux of neutrons produced in a reactor is absorbed by the shielding material, which, in turn, becomes radioactive. Control rods are extremely radioactive; some tritium is produced in water, especially when heavy water is used as a moderator. In general, the lifetime of activated materials can be kept low by selecting low-A material and special steels for infrastructures. The third and most important type of waste is made of *actinides*, i.e. nuclides with $Z \geq 89$. They are produced by neutron capture by ^{238}U, which is the major component of fuel rods. About 0.5% of spent fuel is fissile ^{239}Pu, which is mostly used in nuclear warheads. An example of a nuclide which is present with a concentration of about 0.4% in spent fuel is ^{236}U. This is not present in nature; it has a lifetime of 2.3×10^7 years, and it is at the start of a chain of ten decays, before ending as stable lead:

$$^{236}\text{U} \xrightarrow{\alpha} {}^{232}\text{Th} \xrightarrow{\alpha} {}^{228}\text{Ra} \xrightarrow{\beta} {}^{228}\text{Ac} \xrightarrow{\beta} {}^{228}\text{Th} \xrightarrow{\beta} {}^{224}\text{Ra} \xrightarrow{\alpha}$$

$$\rightarrow {}^{220}\text{Rn} \xrightarrow{\alpha} {}^{216}\text{Po} \xrightarrow{\alpha} {}^{212}\text{Bi} \xrightarrow{\alpha} {}^{208}\text{Tl} \xrightarrow{\alpha} {}^{208}\text{Pb} \qquad (7.45)$$

About 4×10^8 kg of radioactive spent fuel have been produced worldwide so far, with a production rate of 1×10^8 kg/y.

A set of 17 small natural reactors has been discovered in Gabon, West Africa, near the equator, in a place called Oklo, near Franceville. These reactors operated about 2×10^9 years ago, for about one million years. They have been discovered because the uranium ore of these mines has a lower concentration of the fissile isotope.

7.13   The Thorium Cycle

The shortage of Uranium in some countries has revived interest in the thorium cycle, which was abandoned in the USA in 1973. ^{232}Th is the only naturally occurring isotope, for all practical purposes. It is not stable, but decays α with a half-life of 1.4×10^{10} years. It is also not fissile, but it is *fertile*: it can be used to "easily" generate an isotope of uranium, which is fissile, by neutron absorption:

$$n + {}^{232}\text{Th} \rightarrow {}^{233}\text{Th} \rightarrow {}^{233}\text{Pa} \rightarrow {}^{233}\text{U} \qquad (7.46)$$

All the decays in the chain above (Eq. (7.46)) are β^-. The nuclide ^{233}U is fissile. In order to take place, this reaction needs some initial nuclear chain reaction to take place and provide the neutrons to obtain the fissile nuclides. This is an example of a so-called *breeder reactor*, which produces both energy and nuclear fuel. The advantages of thorium are that it is more abundant than uranium, and it is not as diluted.

Also, the spent fuel has a shorter average lifetime, with respect to uranium-based fuel, although some long-lived nuclides are also produced. The only two active thorium-operated research reactors are in India. A recent project, the "energy amplifier", proposes to use a particle accelerator to sustain the criticality condition to an otherwise sub-critical thorium-based reactor, as sketched in Fig. 7.20. One of the advantages is the possibility to operate without initial fissile Pu, which is needed in all other breeder reactors (Eq. (7.43)), and the other is the possibility to fission the actinides.

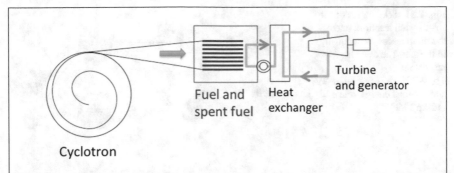

Fig. 7.20 Schematics of the principle of an Energy Amplifier. A proton beam is used as a spallation neutron source to sustain the fission reaction in thorium. This method can also be used to burn actinides from spent nuclear fuel. This generator is at the stage of feasibility study

7.14 Nuclear Fusion and Nucleosynthesis

Nuclear fusion powers the only "controlled" fusion nuclear power source so far: the sun, and other stars. Several fusion reactions occur in sequence in the sun and many have *branches*, which means that they can occur in more than one way. The predominant reaction is called the *proton–proton cycle*, or *pp*-I: protons fuse together to produce deuterons. Other branches of this cycle are also active (*pp*-II and *pp*-III). A star's temperature is much larger than the ionisation temperature of all elements: the sun is made by a neutral plasma of electrons and nuclei. The gravitational force holds them together, creating an enormous pressure. These two conditions make it possible that nuclei get in close contact, penetrating the electrostatic shield that surrounds them. The reactions in the proton–proton (I) cycle are (Fig. 7.21):

$$p + p \rightarrow\ ^2\mathrm{H} + e^+ + \nu_e \quad Q = 0.42\ \mathrm{MeV}\ \text{proton–proton fusion} \tag{7.47}$$

$$p +\ ^2\mathrm{H} \rightarrow\ ^3\mathrm{He} + \gamma \quad Q = 5.49\ \mathrm{MeV}\ \text{proton–deuteron fusion} \tag{7.48}$$

$$^3\mathrm{He} +\ ^3\mathrm{He} \rightarrow\ ^4\mathrm{He} + 2p \quad Q = 12.86\ \mathrm{MeV}\ \text{Helium–Helium fusion} \tag{7.49}$$

Another fusion cycle is very important for stellar evolution: the *carbon-catalysed cycle* (CNO). Carbon acts as a catalyst, meaning that it facilitates the reaction, but carbon nuclei are not created or destroyed by the chain of reactions (Fig. 7.22):

$$p +\ ^{12}\mathrm{C} \rightarrow\ ^{13}\mathrm{N} + \gamma \quad Q = 1.95\ \mathrm{MeV}\ \text{proton–carbon fusion} \tag{7.50}$$

$$^{13}\mathrm{N} \rightarrow\ ^{13}\mathrm{C} + e^+ + \nu_e \quad Q = 1.20\ \mathrm{MeV}\ \beta\text{-decay} \tag{7.51}$$

Fig. 7.21 Schematics of the
pp-I fusion reaction, one of
the main reactions occurring
in the sun. The overall
reaction is
$4p \rightarrow^4 He + 2e^+ + 2\nu_e + 2\gamma$,
and the total energy released
is 24.67 MeV

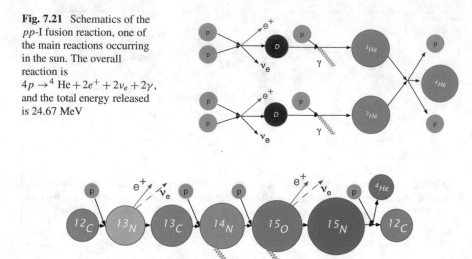

Fig. 7.22 The carbon-catalysed fusion reaction cycle, CNO. Carbon acts as a catalyst

$$p + {}^{13}C \rightarrow {}^{14}N + \gamma \quad Q = 7.55 \text{ MeV proton–carbon} \tag{7.52}$$

$$p + {}^{14}N \rightarrow {}^{15}O + \gamma \quad Q = 7.34 \text{ MeV proton–nitrogen} \tag{7.53}$$

$$^{15}O \rightarrow {}^{15}N + e^+ + \nu_e \quad Q = 1.68 \text{ MeV } \beta\text{-decay} \tag{7.54}$$

$$p + {}^{15}N \rightarrow {}^{12}C + {}^4He \quad Q = 4.96 \text{ MeV proton–nitrogen} \tag{7.55}$$

Carbon takes part in the process; it is consumed, but it is also re-generated, so we find the same amount of carbon in the final state as in the initial state. The overall reaction is

$$4p \rightarrow {}^4He + 2e^+ + 2\nu_e + 3\gamma \quad Q = 24.68 \text{ MeV} \tag{7.56}$$

From all the above reactions, we see that the sun is a large source of electron neutrinos, which are called *solar neutrinos*. Their energy has a range that extends up to 0.4 MeV for *pp*-I reactions and from 1 to 12 MeV, for the CNO reactions. Neutrinos of this energy have a cross section of the order of $\sigma_\nu \approx 10^{-44}$ cm^2. They escape from the sun, which is quite transparent to them, and relatively few of them, for purely geometrical reasons, reach the planet Earth and only very occasionally do they interact with matter, sometimes with neutrino detectors in underground laboratories. By *nucleosynthesis*, we mean the process of formation of elements in stars. When the proton content of stars, including the sun, is depleted, the fusion continues with other reactions, synthetising heavier elements. The sun is a *yellow dwarf star* and when its hydrogen will be consumed, in about 5×10^6 years, it will become a *red giant* star, with a much larger diameter and a core which will start burning helium, to form beryllium, carbon and oxygen; it will not go on further, because the temperature and pressure will not be enough to ignite other fusion reactions: it will become a cold *white dwarf* star. Stars which are more massive continue their fusion processes, which provides less and less energy as heavier nuclides are used for fusion, till producing nuclides with $A \approx 56$. This is the maximum of binding energy per nucleon, as shown in Fig. 7.3; above this nuclear size, adding nucleons would decrease the binding energy and would not be energetically favourable ($Q < 0$). Therefore, heavier elements cannot be produced with normal fusion processes. When a star of a large size, about 10 times the sun mass, has used all its fuel available to fusion, it ends up as an iron core and no radiation to balance the gravitational attraction, which heats up the remaining material and ignites a supernova explosion. Temperature increases above 10^{10} K, and photons have enough energy to break iron nuclei, for $E_\gamma > 2.5$ MeV

$$\gamma + {}^{56}\text{Fe} \rightarrow 13\,{}^{4}\text{He} + 4n \quad (Q = 145 \text{ MeV}) \tag{7.57}$$

The process can continue further, to break alpha particles and create a hot plasma of protons and electrons, which have enough energy to produce an *inverse beta decay*

$$e^- + p \rightarrow n + \nu_e \tag{7.58}$$

What is left is similar to a giant nucleus, with $A \approx 10^{57}$, with a radius of a few kilometres and most of its nucleons in the form of neutrons (a *neutron star*). Under certain conditions, a *supernova* explosion (type-II) takes place, which produces a shock wave of neutrons and neutrinos. Models of supernova explosions predict short time intervals during which emission of a very large neutron flux takes place. This process forms heavier elements with two mechanisms: a sequence of neutron captures and beta decays, starting from initial iron nuclei, or a rapid aggregation of neutrons around an Fe nucleus, and subsequent beta decays. Emission of an even higher flux of neutrinos of all three kinds is also predicted by some theoretical models, and was detected in the laboratory in the last supernova explosion which occurred in proximity of our galaxy. The sun and the solar system have emerged from the remnants of a supernova explosion. The abundance of elements in the solar system is shown in Fig. 7.23.

Some of the solar neutrinos reach our planet, with a flux of about $\Phi_\nu \approx 6.5 \times 10^{10}$ cm^{-2} s^{-1}. A very small fraction of them are detected in underground laboratories. Some of the electron neutrinos transform into μ and τ neutrinos during their flight. The long standing problem of missing solar neutrino flux is now understood in terms of this phenomenon, called *neutrino oscillations*. The solar neutrino problem was first observed by Ray Davis (USA, 1914–2006) Nobel Prize 2002, using the Standard Solar Model calculations by John N. Bahcall (USA, 1934–2005).

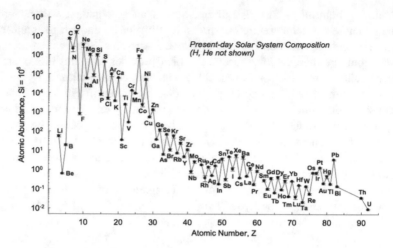

Fig. 7.23 Abundance of elements in the solar system. *From K. Lodders, in "Principles and perspectives in Cosmochemistry", Springer (2010)*

On 23 February 1987, a supernova appeared in the southern hemisphere. The originating star was a blue super giant with originally 10–100 solar masses located in the Large Magellanic Cloud, at "only" 167 kly (thousands of light years) from the Earth. A burst of neutrinos was detected in three neutrinos observatories, at 7 : 35 : 35 universal time, about 3 h before the visible light reached the Earth, which is consistent with a flux of 10^{58} neutrinos in two pulses a few seconds apart. The expected remnant neutron star has not been observed yet.

7.15 Fusion Reactors

The induced nuclear fusion has been achieved on the Earth, but not in a sustained way: in thermonuclear bombs the process is not under control, while in reactors fusion only occurs for a very short time and, so far, not in a self-sustained reaction. The main reactions are

$$^2\text{H} +^2 \text{H} \rightarrow ^3 \text{He} + n; \quad Q = 3.27\,\text{MeV Helium production} \tag{7.59}$$

$$^2\text{H} +^2 \text{H} \rightarrow ^3 \text{H} + p; \quad Q = 4.03\,\text{MeV Tritium production} \tag{7.60}$$

$$^2\text{H} +^3 \text{H} \rightarrow ^4 \text{He} + n; \quad Q = 17.62\,\text{MeV Helium production} \tag{7.61}$$

The main problem is to beat the Coulomb repulsion between protons or tritium nuclei. Magnetic confinement or laser-induced pressure is being used (separately) to reach nuclear fusion in a laboratory. A laser-induced fusion lab in Livermore, CA, has recently reached a breakthrough point: more energy was produced by the fusion process than the energy absorbed by the fuel target. However, this quantity is far less than the energy needed to operate the lasers.

7.16 Problems

For these problems, we need to use more decimal digits than usual, because some of them are based on detailed calculations, involving differences between quantities which differ by less than 1%. Some of the data needed is available in the text.

7.1 The deuteron is a (pn) bound state. Its *atomic* mass is 2.014101 a.m.u. (or u). Calculate its binding energy. The mass of the proton is 1.0072764 u and the mass of the neutron is 1.008665 u.

7.2 Repeat the calculation above using MeV/c^2 for masses. The value of the masses are: deuteron: 1875.613 MeV/c^2; proton: 938.272 MeV/c^2; neutron: 939.565 MeV/c^2.

7.3 The atomic mass of $^{241}_{95}Am$ is 241.056829 u. What is its mass deficit? It decays to $^{237}_{93}Np$ by emitting an α particle. The atomic mass of Np is 237.048173 u. What is the kinetic energy of the α emitted?

7.4 In a fission nuclear reactor, a possible reaction is

$$n +^{235} U \rightarrow^{92}_{37} Rb +^{140}_{55} Cs + N \text{ neutrons}$$

How many neutrons are produced? (the baryon number is conserved).

7.5 In the above reaction, the atomic masses are

$$M(^{235}_{92}U) = 235.04393 \text{ u}$$

$$M(^{92}_{37}Rb) = 91.919729 \text{ u}$$

$$M(^{140}_{55}Cs) = 139.91728 \text{ u}$$

$$M(n) = 1.008665 \text{ u}$$

Neglecting the kinetic energy of the incoming neutron, is the reaction endo-energetic or exo-energetic? What is the Q-value? (in MeV).

7.6 ^{232}Th is the only naturally occurring isotope of thorium. Its half-life is 14.05×10^9 years. What is its mean lifetime in seconds? A mole of thorium (232 g) is made of 6.02×10^{23} atoms (Avogadro's number).

(a) What is the activity of a mole of thorium, in Bq ($[s^{-1}]$)?

(b) The decay reaction is ^{232}Th $\rightarrow$ ^{228} Ra $+ \alpha$ with a kinetic energy of the
α-particle of 4.083 MeV. Suppose that all the radiation is absorbed by the
brick containing our thorium sample and transformed into heat. How much
energy per second (power) is generated?

7.17   Solutions

Solution to 7.1   As for all nuclides, the mass of the deuteron is smaller than the
sum of the mass of its components. The binding energy is equal to this mass
difference. In this calculation, care must be taken to include or not the electron
mass. This is normally included in the value of the *atomic mass*, but not in
the proton mass, even if it is expressed in units of u. The electron mass is
510.9989 keV/c^2, $u = 931,494.0$ keV/c^2, so $m_e = 0.0005485\, u$.

$$\Delta M = M_D - m_e - m_p - m_n =$$

$$= 2.0141018 - 0.0005485 - 1.0072764 - 1.0086649 = -0.0023880\, \text{u}$$

Expressing it in MeV/c^2, we have

$$B = 931.4940 \times 0.0023880 = 2.2244\, \text{MeV, or } 1.1122\, \text{MeV per nucleon.}$$

Solution to 7.2   When masses are given in MeV, normally electrons are not
included. In the data of this problem, it is specified that the mass of the *deuteron*
is given. The calculation is straightforward:

$$\Delta M = m_D - m_p - m_n\, ;$$

$$\Delta M = 1875.613 - 938.272 - 939.565 = -2.224\, \text{MeV/c}^2$$

Solution to 7.3   The Am isotope has 95 protons and 146 neutrons. Its mass
deficit is

$$\Delta M = M_{\text{Am}} = 241.056827 - 95 \times (1.0072764 + 0.0005485) - 146 \times 1.008665$$

$$\Delta M = -1.9516285\, \text{u}$$

$$B = -\Delta M c^2 = 1.9516285 \times 931.4940 = 1817.93\, \text{MeV} = 7.543\, \text{MeV per nucleon}$$

The Q-value of the reaction

$$^{241}_{95}\text{Am} \rightarrow \alpha +^{237}_{93} \text{Np} + 2\, e^- \qquad \text{is}$$

$$Q = M_{\text{Am}} - M_{\text{Am}} - M_\alpha - 2m_e$$

$$Q = 241.056827 - 237.048173 - 4.0015061 - 2 \times 0.0005485 = 0.006050 \text{ u}$$

which translates into 5.636 MeV. The largest part of this energy is converted into kinetic energy of the alpha particle, recoiling against a ^{237}Np nucleus: $E_k(\alpha) = 5.485$ MeV.
Solution to 7.4

$$n +{}^{235}\text{U} \rightarrow {}^{92}_{37}\text{Rb} + {}^{140}_{55}\text{Cs} + N \text{ neutrons}$$

Uranium has 92 protons. The number of protons and neutrons is conserved.
Protons: $92 = 37 + 55$
Neutrons: $143 + 1 = 55 + 85 + N \Rightarrow N = 4$
Solution to 7.5

$$Q = 1.008665 + 235.04393 - 91.919729 - 139.91728 - 4 \times 1.008665 = +0.180926$$

This value is positive, the reaction is exo-energetic; it releases energy.
Solution to 7.6

$$\tau = 14.05 \times 10^9 \times 3.1536 \times 10^7 = 44.3 \times 10^{16} \, s$$

$$A = N/\tau = 6.02 \times 10^{23}/44.3 \times 10^{16} = 1.36 \times 10^6 \text{ Bq}$$

$$P = 1.36 \times 10^6 \times 4.083 \times 10^6 \times 1.602 \times 10^{-19} \text{ W}$$

$$P = 8.89 \times 10^{-7} W$$

Bibliography and Further Reading

B.R. Martin, *Nuclear and Particle Physics - An Introduction*, 2nd edn. (Wiley, Hoboken, 2009)
E. Mervine, *Nature's Nuclear Reactors: The 2-Billion-Year-Old Natural Fission Reactors in Gabon, Western Africa* , Scientific American July 13, 2011
Y. Oka, *Nuclear Reactor Design* (Springer, Tokyo, 2014)
D. Perkins, *Particle Astrophysics* (Oxford University Press, Oxford, 2009)
P. Schuck, A. Toshaki et al., Alpha-particle condensation in nuclear systems: present status and perspectives. J. Phys. Conf. Ser. 436 (2013)
W.M. Stacey, *Nuclear Reactor Physics* (Wiley, Hoboken, 2007)
T. Yamada et al., Nuclear alpha-particle condensates, in *Clusters in Nuclei*, vol. 2, ed. by C. Beck. Lecture Notes in Physics, vol. 848 (Springer, 2012), p. 229

Chapter 8
Six Problems

8.1 Voyager Power Source

The space probe Voyager-2 (Fig. 8.1) was launched on August 20, 1977 and is still sending data to the Earth, from a distance of 1.6×10^{10} km.

The instruments on board are powered by a thermoelectric generator, which produces electrical power from the heat generated by the decay of radioactive nuclides. In this particular case, the parent nuclide is ^{238}Pu. The half-life of this nuclide is $t_{1/2} = 87.7$ years. We can consider the electric power as proportional to the activity of the radioactive source. The initial electric power at launch was 420 W. What is the electric power now, and at what date will the space probe generator deliver only 10% of the initial electrical power?

Solution
First of all, we should convert the half-life into mean-lifetime:

$$\tau = \frac{t_{1/2}}{\ln 2} = t_{1/2} \times 1.443 = 87.7 \times 1.443 = 126.52 \text{ years}$$

If you don't remember this formula, you can calculate it quickly:

$$\frac{N(t)}{N_0} = \frac{1}{2} = e^{-\frac{t_{1/2}}{\tau}}$$

Taking logs ($\ln = \log_e$) on both sides:

$$\ln(1) - \ln 2 = -\frac{t_{1/2}}{\tau}$$

$$\tau = \frac{t_{1/2}}{\ln 2} = \frac{87.7}{0.693} = 126.5 \text{ years}$$

© Springer Nature Switzerland AG 2018
S. D'Auria, *Introduction to Nuclear and Particle Physics*,
Undergraduate Lecture Notes in Physics,
https://doi.org/10.1007/978-3-319-93855-4_8

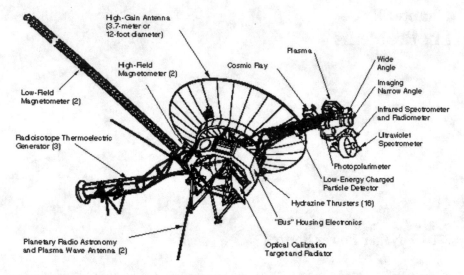

High-Gain Antenna
(3.7-meter or
12-foot diameter)

Plasma

High-Field
Magnetometer (2)

Cosmic Ray

Wide
Angle

Low-Field
Magnetometer (2)

Imaging
Narrow Angle

Infrared Spectrometer
and Radiometer

Radioisotope Thermoelectric
Generator (3)

Ultraviolet
Spectrometer

Photopolarimeter

Low-Energy Charged
Particle Detector

Hydrazine Thrusters (16)

"Bus" Housing Electronics

Planetary Radio Astronomy
and Plasma Wave Antenna (2)

Optical Calibration
Target and Radiator

Fig. 8.1 The Voyager-2 space probe (NASA)

The mean lifetime is always larger than the half-life by a factor 1.4 (or $\ln 2 = 0.693 < 1$). Now, you can calculate the electric power W as a function of time. Let's call $\mathcal{A}(t)$ the activity, we have $W(t) = K\mathcal{A}(t)$, where K is a constant with the dimension of an energy. We can write this because we know that the power is proportional to the activity.

$$W(t) = K\mathcal{A}(t) = K\frac{N(t)}{\tau} = \frac{K}{\tau}N_0 e^{-t/\tau}$$

We realise that

$$\frac{K}{\tau}N_0 \text{ is simply the initial power } W_0$$

$$W(t) = W_0 e^{-t/\tau}$$

Numerical calculation: 37 years have passed since the launch, so the power is reduced by a factor $\exp(-37/126.5) \approx 75\%$.

These days the Voyager-2 space probe has a 313-W power generator.

$$\frac{W(t)}{W_0} = 0.10 = e^{-\frac{t_{10\%}}{\tau}}$$

$$-\ln 10 = -\frac{t_{10\%}}{\tau}; \quad t_{10\%} = \tau \ln 10 = 126.52 \times 2.30 = 291.3 \text{ years}$$

from the launch, which means that Voyager-2 will have 10% of its original power in $1977 + 291 = 2268$ a.D.

8.2 Plutonium Thermoelectric Generator

The nuclide ^{238}Pu decays α with a half-life $t_{1/2} = 87.7$ years. What is the Q-value of the decay and what is the corresponding thermal power per kilogram? The corresponding isotopic masses (in atomic mass unit (u) or Dalton) are

$M(^{238}\text{Pu}) = 238.04956$ u;
$M(^{234}\text{U}) = 234.04095$ u;
$M(\alpha) = 4.00151$ u;
$1u = 931.494$ MeV/c^2
1 MeV $= 1.6 \times 10^{-13}$ J;
1 year $= 3.15 \times 10^7$ s;
$N_A = 6.022 \times 10^{23}$ mol^{-1};

Solution

From the definition of Q-value

$$Q = E_k(\text{final particles}) - E_k(\text{initial particles})$$

we derive

$$Q = \overset{\text{initial}}{\sum M_i} - \overset{\text{final}}{\sum M_f}$$

that is, it is the sum of the masses in the initial state minus the sum of the masses of the final state.

$$Q = M(^{238}\text{Pu}) - M(^{234}\text{U}) - M(\alpha)$$

In 1 kg of plutonium-238, there are $N_0 = \frac{1000\text{g}}{238\text{g}} N_A$ nuclei.

The initial activity is $\mathcal{A} = N_0/\tau$, where $\tau = t_{1/2}/\ln 2$.

Assuming that all energy is converted into heat, the thermal power radiated by 1 kg of Plutonium is

$$W = Q\mathcal{A} = Q \frac{1000\,\text{g}}{238\,\text{g}} N_A \frac{\ln 2}{t_{1/2}}$$

Explanation: Each decay releases an energy Q; we have $\mathcal{A}$ decays per second, so the energy released per second (which is the required power) is $W = Q\mathcal{A}$.

Numerical calculations:

$$Q = M(^{238}\text{Pu}) - M(^{234}\text{U}) - M(\alpha)$$

$$= (238.04956 - 234.04095 - 4.00151) \times 931.494 = 6.6\,\text{MeV}$$

$$W = 6.6\,\text{MeV} \times 1.6 \times 10^{-13}\,\text{J/MeV}\,\frac{1000\text{g}}{238\text{g}} \times 6.022 \times 10^{23}\frac{\ln 2}{87.7 \times 3.15 \times 10^7\text{s}}$$

$$= 0.67 \times 10^3\,\text{W}$$

If not adequately cooled, plutonium is intrinsically quite hot, producing 670 W/kg. Incidentally, 1 kg of plutonium has an activity of 63×10^{12} Bq = 63 TBq.

8.3   Alpha and Beta Decays

1. The nuclide ^{238}Pu decays α. What is the mass number of the daughter nuclide?
2. The nuclide $^{102}_{45}\text{Rh}$ (rhodium) can decay to β^+ (80%) and β^- (20%). (a) What are the mass and atomic numbers (A and Z) of the two possible daughters? (b) What decay process is this?
3. The nuclide $^{90}_{38}\text{Sr}$ (strontium) decays β^- to another radioactive nuclide which also decays β^-. (a) Write the mass and atomic numbers (A and Z) of both the daughter and grand-daughter nuclides. (b) What decay process is this? (c) Which fundamental interaction is mediating this decay?

Solutions:

1. In α decays $A \rightarrow A' = A - 4$ and $Z \rightarrow Z' = Z - 2$ because an alpha particle ($A = 4$, $Z = 2$) is emitted. In our case, the decay is

$$^{238}\text{Pu} \rightarrow\,^{234}\text{U} + \alpha(+5.593\text{MeV})$$

2. Beta decays leave A unchanged.
   In case of β^+, a positron is emitted, so the number of protons decreases by one: $Z \rightarrow Z' = Z - 1$.
   In case of β^-, an electron is emitted, so the number of protons increases by one: $Z \rightarrow Z' = Z + 1$.
   In our case:

$$^{102}_{45}\text{Rh} \rightarrow\,^{102}_{46}\text{Pd} + e^- + \bar{\nu}_e \quad (A' = 102,\, Z' = 46)$$

$$^{102}_{45}\text{Rh} \rightarrow\,^{102}_{44}\text{Ru} + e^+ + \nu_e \quad (A' = 102,\, Z' = 44)$$

(b) this decay process is a "branching".

3. (a)

$$^{90}_{38}\text{Sr} \rightarrow ^{90}_{39}\text{Y} \rightarrow ^{90}_{40}\text{Zr} \quad (A' = 90, \ Z' = 39; \ A'' = 90, \ Z'' = 40)$$

(b) This is a chain of β decays.
(c) The beta decays are due to the weak part of the electroweak interaction.

8.4 Relativistic Muons

The purpose of these two exercises is to make you familiar with calculations involving relativistic formulae, in particular how to derive the γ factors from momentum and kinetic energy of the particle. Muons are produced by cosmic rays when colliding mostly with nitrogen nuclei in the upper atmosphere. Calculate the mean life of a muon with an initial kinetic energy $E_K = 20$ GeV in the "laboratory" reference frame. Calculate also its velocity $\beta = v/c$. Having such a velocity, the muon mean lifetime corresponds to a mean decay length. This means that if a set of 20 GeV muons are all produced at a point A, and each of them decays at a point $D_1, D_2, D_3 \ldots$ we can calculate an average distance AD_{ave} from the average lifetime and the muon velocity. Compare this length with the top height of the stratosphere (50 km). The muon mass is $m_\mu = 105.66$ MeV/c^2, and the muon lifetime (at rest) is $\tau_\mu = 2.20\,\mu$s.

Solution
We can calculate γ starting from the definition of relativistic kinetic energy of a particle :

$$E_k = (\gamma - 1)mc^2$$

$$\gamma = 1 + \frac{E_k}{mc^2} = 1 + \frac{20000\,(\text{MeV})}{105.66\,(\text{MeV/c}^2)c^2} = 190.3$$

The muon lifetime, when observed from a reference frame where the muon is in motion, is larger by the factor γ related to the velocity of the muon:

$$\tau_{(\text{lab})} = \gamma\tau = 190.3 \times 2.20\,\mu\text{s} = 418.6\,\mu\text{s}$$

From the definition of γ:

$$\gamma = \frac{1}{\sqrt{1 - \beta^2}}$$

We can square both sides, to obtain

$$\gamma^2 \left(1 - \beta^2\right) = 1; \; \beta^2 = 1 - \frac{1}{\gamma^2}$$

$$\beta = \sqrt{1 - \frac{1}{\gamma^2}} = \sqrt{1 - \frac{1}{190.3^2}} = 0.9999862$$

We can approximate this velocity with $c \approx 3 \times 10^8$ m/s.

The mean decay length of a muon with a given velocity is the length traveled in the mean lifetime:

$$c\tau_{(\text{lab})} = 3 \times 10^8 (\text{m/s}) \times 418.6 \times 10^{-6} (\text{s}) = 125589 (\text{m}) = 125\,\text{km}$$

This value is about twice the height of the top of the stratosphere: energetic muons can travel all the way down to ground before decaying.

8.5   Same Exercise, Starting from the Muon Momentum

In some cases, it is easier to measure the momentum of a charged particle, by measuring the bending radius of its trajectory in a magnetic field. Suppose that we measure the momentum p of the muon to be 20 GeV/c. What is the corresponding gamma factor? All data are given in the problem above.

Solution

There are at least two ways to solve this: (a) to calculate the muon kinetic energy from its momentum, and then following the previous case, or (b) calculate $\beta\gamma$ from the definition of relativistic momentum.

Method (a):

From the modulus of the muon 4-momentum $E^2/c^2 - p^2 = m_\mu^2 c^2$, we derive

$$E = \sqrt{m^2 c^4 + p^2 c^2}; \; \text{This is total energy } E = E_k + mc^2$$

$$E_k = E - mc^2 = \sqrt{m^2 c^4 + p^2 c^2} - mc^2$$

$$= \sqrt{105.66^2 + 20000^2} - 105.66 = 19.894\,\text{GeV}$$

We can continue just as in the problem above: from the definition of relativistic kinetic energy, we can calculate γ:

$$E_k = (\gamma - 1)mc^2$$

$$\gamma = 1 + \frac{E_k}{mc^2} = 1 + \frac{19{,}894 \ (\text{MeV})}{105.66 \ (\text{MeV}/c^2)c^2} = 189.3$$

and so on. Numerically, this value is not so different from the value calculated from a kinetic energy of the same value as the momentum. As long as the energy or the momentum are much larger than the mass, which can be expressed in terms of energy, in MeV/c^2, momentum and kinetic energy are good approximations of each other.

Method (b)

The momentum of the muon along its flight direction is $p = \beta\gamma c m_\mu$ so that

$$\beta\gamma = \frac{p}{c m_\mu} = \frac{20000 \ (\text{MeV}/c)}{105.66 \ (\text{MeV}/c^2) \ c} = 189.3 \qquad (8.1)$$

We can now calculate β^2 from γ^2

$$\gamma^2 = \frac{1}{(1 - \beta^2)} \rightarrow 1 - \beta^2 = \frac{1}{\gamma^2}$$

So, $\beta^2 = 1 - \frac{1}{\gamma^2}$. Therefore:

$$\beta^2\gamma^2 = \left(1 - \frac{1}{\gamma^2}\right)\gamma^2 = \gamma^2 - 1$$

Now from (Eq. (8.1)), we can square both sides and obtain

$$\beta^2\gamma^2 = \frac{p^2}{m^2c^2} = \gamma^2 - 1, \text{ where the last step comes from the equation above.}$$

$$\gamma = \sqrt{1 + \frac{p^2}{m^2c^2}} = \sqrt{1 + \frac{20000^2(\text{MeV}^2/c^2)}{105.66^2 \ (\text{MeV}^2/c^4)c^2}} = 189.3 \ .$$

Comparing this value with the value of $\beta\gamma$ obtained in (Eq. (8.1)), we can say that $\beta \approx 1$ and $v \approx c$.

Also in this case, the proper lifetime is increased by a factor γ:

$$\tau_{(\text{lab})} = \gamma\tau_\mu = 189.3 \times 2.20 = 416.4\mu s \ .$$

The mean decay length is therefore:

$$l = \beta c \tau (\text{lab}) = 3 \times 10^8 \text{ m/s} \times 416.4 \times 10^{-6} \text{ s} = 124.9 \times 10^3 \text{ m}.$$

8.6 Chernobyl Caesium Release

In April 1985, one of the nuclear reactors in a power plant in Ukraine released, in total, about 26 kg of ^{137}Cs, among many other radioactive nuclides. This was the worst accident in the history of nuclear power plants. This radionuclide decays β^- with half-life of 30.17 years. Calculate the total radioactivity (in Bq = $[s^{-1}]$) from this nuclide only, which was released in the environment by the accident. Calculate also the total Chernobyl-related ^{137}Cs residual activity which is still present today. A solution is given for year 2014. One year is 3.15×10^7 s, and Avogadro's number is $N_A = 6.022 \times 10^{23}$ mol^{-1};

Solution

The ^{137}Cs lifetime is

$$\tau = t_{1/2}/\ln 2 = 30.17 \times 1.44 \times 3.15 \times 10^7 = 137 \times 10^7 \text{s}$$

We need to know how many Cs atoms were released. So, we first need to know how many moles correspond to $M = 26$ kg, then multiply by the number of atoms per mole, which is Avogadro's number, and divide by the lifetime to obtain the activity.

The released activity is

$$A = \frac{N_0}{\tau} = \frac{M}{M_A} \frac{N_A}{\tau} = \frac{26{,}000(\text{g})}{137(\text{g/mol})} \frac{6.022 \times 10^{23} \text{ /mol}}{137 \times 10^7(\text{s})} = 83 \times 10^{15} \text{ Bq}$$

The disaster released 83 PBq of ^{137}Cs. 29 years later the residual ^{137}Cs activity is

$$A(t) = A_0 e^{-\frac{t}{\tau}} = 83 \times 10^{15} e^{-\frac{29(y)}{1.44 \times 30.17(y)}} = 83 \times 10^{15} \times 0.512 = 42 \text{ PBq}$$

About 29 years later, the caesium activity is a little more than half the released activity. The disaster occurred about one caesium half-life from 2014.

Name Index

© Springer Nature Switzerland AG 2018
S. D'Auria, *Introduction to Nuclear and Particle Physics*,
Undergraduate Lecture Notes in Physics,
https://doi.org/10.1007/978-3-319-93855-4

Subject Index

A

Actinides, 171–173
Activity, 2, 3, 31, 54, 55, 61, 64, 65, 67, 68, 171, 177, 181–184, 188
Asymptotic freedom, 144
Atomic density, 73, 83
Atomic mass unit, 150, 183
Attenuation coefficient, 74, 77, 78, 98–100

B

Beta decays, 26, 52, 56, 66, 92, 128, 147, 153–156, 165, 175, 184–185
Bethe–Bloch formula, 82, 84, 168
Binding energy, 79, 151–153, 159, 161, 164, 166, 175, 177, 178
Bosons, 26, 104, 105, 107, 113, 116–121, 124, 126, 127, 129–132, 162
Branching fraction, 58, 120, 145, 146, 163
Breeder reactor, 170, 172
Bremsstrahlung, 88–91, 97, 99, 102

C

Carbon-catalysed cycle, 173, 174
Chain reaction, 168, 172
Charm quark, 137
Collective model, 157, 161
Condensates, 161, 179
Confinement, 26, 140, 141, 177
Criticality, 167, 168, 172

D

Dalton (unit), 108, 183
Decay width, 106–108

D

Decuplet, 139, 140
Detailed balance, 109, 110
Deuteron, 46, 147, 148, 173, 177, 178
Direct product, 30, 31, 138
Direct sum, 30, 31, 137
Doppler effect, 26–27, 33, 35

E

Elastic scattering, 149, 164–165, 168
Electron capture, 63, 154–156
Energy amplifier, 173
Equivalent dose rate, 95
Evaporation, 160
Exclusion principle, 6, 116–118, 157

F

Fermi gas nuclear model, 157–159
Fermi level, 158
Fermions, 116–118, 120–123, 125–127, 131, 156
Fine structure constant, 82, 109, 113
First integrals, 30
Flavour, 122, 124, 125, 130, 134–137, 139–142, 145

G

Gamma emission, 162–163
Gluon, 26, 105, 119, 120, 126, 127, 129, 131–133, 142–144
Golden rule, 109
Gravity, 1, 103, 105, 145, 151
Group theory, 28–30, 45, 134–136

© Springer Nature Switzerland AG 2018
S. D'Auria, *Introduction to Nuclear and Particle Physics*,
Undergraduate Lecture Notes in Physics,
https://doi.org/10.1007/978-3-319-93855-4

Printed in the United States
By Bookmasters